Adaptive Digital Circuits for Power-Performance
Range beyond Wide Voltage Scaling

Saurabh Jain • Longyang Lin • Massimo Alioto

# Adaptive Digital Circuits for Power-Performance Range beyond Wide Voltage Scaling

From the Clock Path to the Data Path

 Springer

Saurabh Jain
National University of Singapore
Singapore, Singapore

Massimo Alioto
National University of Singapore
Singapore, Singapore

Longyang Lin
National University of Singapore
Singapore, Singapore

ISBN 978-3-030-38798-3        ISBN 978-3-030-38796-9    (eBook)
https://doi.org/10.1007/978-3-030-38796-9

This Springer imprint is published by the registered company Springer Nature Switzerland AG
The registered company address is: Gewerbestrasse 11, 6330 Cham, Switzerland

*To Sarakshi, my daughter and my constant source of smile.*

*To my family, relatives, and friends.*

*To Maria Daniela, Marco and Marina, and all my nephews and nieces including Flavio now. And to the generous relatives and friends who have encouraged my growth and inspired true enjoyment in the pursuit of excellence.*

# Preface

A plethora of new applications are demanding the availability of digital sub-systems with an uncommonly wide power-performance range, especially in systems that are tightly power-constrained (e.g., battery-powered, energy harvested). Examples of such applications are intelligent IoT sensor nodes, wearables, and biomedical devices with on-chip event detection and data analytics capabilities, intelligent and reliable components for automotive with built-in data analysis, always-on computer vision, always-on systems with on-chip AI and machine learning, among the many others.

The above applications require more efficient exploitation of the dynamic nature of the workload to reduce the energy consumption during regular operation, while delivering high levels of performance to promptly respond to incoming events when needed. Such demand for wide power-performance range has become common-place in time-driven systems where operation is duty cycled to reduce the average power. Such demand is now becoming pressing in systems that are event-driven to make the power cost of always-on event detection affordable.

As wide voltage scaling has been already exploited for a decade, it is now running out of steam and the expansion of the power-performance range mandates innovation in other dimensions. The limitations of wide voltage scaling in terms of energy and performance gain, voltage scalability, and robustness stem from the high sensitivity of two fundamental design tradeoffs to the adopted supply voltage. Indeed, wide voltage scaling typically alters the ratio of leakage and dynamic energy consumption by one to two orders of magnitude. In turn, this ratio is a crucial parameter that defines the energy-optimal microarchitecture and the logic depth in the data path. The same observation applies to the ratio of the wire and gate delay, which in turn is the key parameter that defines the optimal topology of the clock network that meets the clock skew target. Accordingly, wide voltage scaling makes any fixed data and clock path design sub-optimal at one end of the voltage range, or the other.

The grand goal of this book is to offer the first comprehensive coverage of digital design techniques that expand the power-performance tradeoff well beyond allowed by conventional wide voltage scaling. Reconfiguration in the data and the clock path is introduced to enable dynamic adaptation to the wide change in the leakage–dynamic energy ratio, as well as the wire–gate delay ratio. This circumvents the traditional designer's dilemma of choosing which end of the power-performance spectrum is favored over the other, when adopting wide voltage scaling.

The content of this book is based on our recent research on the topic, as well as the effort to develop innovative design techniques, suitable design methodologies, and silicon prototype demonstrations. To make such design solutions widely usable in our community and the related applications, drop-in techniques have been developed to enable fully automated and low-effort design based on commercial design tools and digital flows. Such techniques have been devised and experimentally validated for the different classes of digital sub-systems, from processors, to accelerators and on-chip memories. To facilitate the usability of the concepts presented in this book, all the scripts necessary to create automated design flows are provided, explained, and publicly shared (see Appendix).

There are many ways to use this book. In particular, it can serve as a reference to practicing engineers working in the broad area of integrated circuit/system design targeting a wide power-performance tradeoff. The book is also very well suited for undergraduate and graduate students; thanks to the rigorous and formal coverage of the underlying topics, and a rich set of references.

Ultimately, we hope that our work will be truly useful to researchers to enable further exploration of reconfigurable microarchitectures and clock networks, as well as to design engineers to make an impact in new product developments. We hope you will enjoy the reading.

Singapore, Singapore                                                                        Saurabh Jain
December 2019                                                                              Longyang Lin
                                                                                          Massimo Alioto

# Acknowledgements

The authors would like to thank

- Intel corp. for funding support, as well as valuable feedback and discussion
- the National Research Foundation (NRF, "CogniVision" grant NRF-CRP20-2017-0003—*www.green-ic.org/cognivision*) for funding support
- Cadence Design Systems, Inc. for EDA tool support
- TSMC for testchip fabrication.

# Contents

# About the Authors

**Saurabh Jain** took his bachelor's and master's degree (a dual degree program) from the Indian Institute of Technology, Kanpur, with specialization in the VLSI field in 2013. He completed his PhD from National University of Singapore in November 2018. His doctoral thesis was on energy efficient reconfigurable microarchitecture development for processors, memory and DSP accelerators. He has worked as a Research Fellow at NUS in ECE department and was involved in SRAM design and development of in-memory solutions for convolutional neural network-based image recognition. He is currently working as a research scientist in the Processor Architecture Research Lab at Intel labs, Bangalore.

**Longyang Lin** received the dual bachelor's degrees from Shenzhen University, Shenzhen, China and Umeå University, Umeå, Sweden, in 2011 and the master's degree from Lund University, Lund, Sweden, in 2013, and the PhD degree from the National University of Singapore, Singapore, in 2018. He is currently a postdoctoral research fellow at the Department of Electrical and Computer Engineering of the National University of Singapore.

His research interests include ultra-low power VLSI circuits, self-powered sensor nodes, widely energy-scalable VLSI circuits and general purpose compute-in-memory.

**Massimo Alioto** received the Laurea (MSc) degree in Electronics Engineering and the PhD degree in Electrical Engineering from the University of Catania (Italy) in 1997 and 2001, and the Bachelor of Music in Jazz Studies from the Conservatory of Music of Bologna in 2007. He is with the Department of Electrical and Computer Engineering, National University of Singapore, where he leads the Green IC group and is the Director of the Integrated Circuits and Embedded Systems area. Previously, he held positions at the University of Siena, Intel Labs—CRL (2013), University of Michigan Ann Arbor (2011–2012), BWRC—University of California, Berkeley (2009–2011), and EPFL (Switzerland, 2007).

He has authored or co-authored more than 280 publications on journals and conference proceedings. He is co-author of other three books, including *Enabling the Internet of Things—from Circuits to Systems* (Springer, 2017), *Flip-Flop Design in Nanometer CMOS—from High Speed to Low Energy* (Springer, 2015), and *Model and Design of Bipolar and MOS Current-Mode Logic: CML, ECL and SCL Digital Circuits* (Springer, 2005). His primary research interests include self-powered wireless integrated systems, near-threshold circuits for green computing, widely energy-scalable and energy-quality scalable integrated systems, data-driven integrated systems, hardware-level security, and emerging technologies, among the others.

He is the Editor in Chief of the IEEE Transactions on VLSI Systems (2019–2020) and was the Deputy Editor in Chief of the IEEE Journal on Emerging and Selected Topics in Circuits and Systems (2018). In 2009–2010, he was Distinguished Lecturer of the IEEE Circuits and Systems Society, for which he is/was also member of the Board of Governors (2015–2020) and Chair of the "VLSI Systems and Applications" Technical Committee (2010–2012). In 2020–2021, he is also Distinguished Lecturer for the IEEE Solid-State Circuits Society. In the last 5 years, he has given 50+ invited talks in top conferences, universities, and leading semiconductor companies. His research has been mentioned in more than 60 press releases and popular science articles in the last 2 years.

He served as Guest Editor of several IEEE journal special issues (e.g., TCAS-I, TCAS-II, JETCAS). He also serves or has served as Associate Editor of a number of IEEE and ACM journals. He is/was Technical Program Chair (ISCAS 2023, SOCC, ICECS, NEWCAS, VARI, ICM, PRIME) and Track Chair in a number of conferences (ICCD, ISCAS, ICECS, VLSI-SoC, APCCAS, ICM). Currently, he is also in the IEEE "Digital Architectures and Systems" ISSCC sub-committee and the IEEE ASSCC technical program committee. Prof. Alioto is an IEEE Fellow.

# Chapter 1
# Introduction

**Abstract** This chapter opens the book and provides a brief analysis of historical trends in digital systems. Highlighting the gradual evolution of the semiconductor industry towards distributed computing (e.g., Internet of things—IoT), the demand for increasingly lower power in the common case and much higher peak performance is discussed. This has led to the introduction of various popular techniques such as wide dynamic voltage frequency scaling (DVFS). Then, the challenges posed by wide DVFS in both the clock and the data path are discussed, motivating the next chapters of the book.

**Keywords** Dynamic voltage frequency scaling (DVFS) · Distributed computing · Internet of things (IoT) · Machine learning · Artificial intelligence (AI) · Energy trends · Cost trends · Semiconductor market · S-curve · Gompertz function · Power budget · Microprocessor trends · Graphics processing unit (GPU) · Digital signal processor (DSP) · Koomey's law · Gene's law · Supercomputer · Short-range radios · Energy per AD conversion · Power-limited systems · Battery-powered systems · Carrier modulation · Time-driven systems · Event-driven systems · Duty cycling · Wake-up · Always-on · Latency · Intelligent sensor nodes · Speech recognition · Smart buildings · Applications · Computer vision · Audio monitoring · Dynamic energy · Leakage energy · Activity factor · Switched capacitance · Supply voltage · Leakage current · Clock cycle · Wire delay · Gate delay · Leakage/dynamic energy ratio · Wire/gate delay ratio

This chapter opens the book and provides a brief analysis of historical trends in digital systems. Highlighting the gradual evolution of the semiconductor industry towards distributed computing (e.g., Internet of things—IoT), the demand for increasingly lower power in the common case and much higher peak performance is discussed. This has led to the introduction of various popular techniques such as wide dynamic voltage frequency scaling (DVFS). Then, the challenges posed by wide DVFS in both the clock and the data path are discussed, motivating the next chapters of the book.

In this book, digital circuit and microarchitectural reconfiguration techniques are introduced to extend the power-performance tradeoff over a range that exceeds conventional wide voltage scaling. Compared to conventional fixed designs, the reconfiguration approaches described in this book make digital circuits more versatile and adaptive, allowing simultaneous optimization at both ends of the power-performance spectrum. Drop-in solutions for fully automated and low-effort design based on commercial design tools are extensively discussed and validated through silicon measurements.

Overall, the main goal is to equip the reader with easy and agile design solutions to generate dynamically adaptable microarchitectures that enhance the benefits of wide DVFS. To enable the reader to quickly utilize the content of the book, automated design flows and the necessary scripts to implement them have been made publicly available.

## 1.1   Trends in the Semiconductor Industry

The semiconductor industry has evolved according to trends that exhibit an exponential improvement over time in many technological and economic respects. As shown in Fig. 1.1, the global sale volume of devices with compute capabilities has increased at a pace of 20× every decade, shaping the semiconductor market size as a succession of superimposed technological waves, each of which can be modeled as a time series following the Gompertz function (often named as "S-curve") [1, 2]. Overall, this has added 11B$ per year to the semiconductor market size on average [1], as shown in Fig. 1.2. The resulting economy of scale and technological advances have led to a form factor shrinkage by nearly two orders of magnitude every decade,

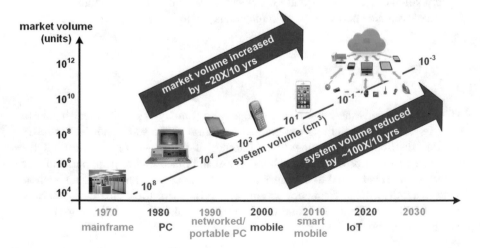

**Fig. 1.1**  Scaling laws of computer market size

leading the centimeter scale in the last decade and leading to millimeter scale in the upcoming one [1, 3–5]. At the same time, 2× reduction in the cost per transistor [1] has made integrated systems more inexpensive and has hence favored relentless proliferation of such small systems. These trends have progressively enabled distributed computing and triggered the technological wave of the Internet of things (IoT) and its flavor with edge computing and AI aboard (see Figs. 1.1 and 1.2).

At the same time, continuous miniaturization and distributed computation have been made possible by relentless reductions in the energy consumption per operation, which in turn have enabled higher performance within the same average power envelope. The latter is generally imposed by the application and the system form factor target as in Fig. 1.3, ranging from nW to µW in distributed sensing platforms with energy harvesting, to W-range in portable and battery-powered electronics, to several Ws in personal computing devices, to tens or few hundreds of Ws in high-performance systems. Detailed historical energy trends for different sub-systems are shown in Fig. 1.4a–d [6]. From Fig. 1.4a, the energy per computation in general-purpose platforms (e.g., microprocessor- and DSP-based) has consistently been decreasing by about 100× every decade and has slowed down to one order of magnitude in recent years. General-purpose GPUs are scaling down by such 10–20× rate as well, suggesting that the "new normal" in computation is really one order of magnitude per decade, with GPUs scaling about twice as fast. Similar observations can be made on the highest end of the spectrum of computers, i.e., super-computers, as shown in Fig. 1.4b. Interestingly, sensor interfaces and analog-to-digital converters have been following a similar trend from Fig. 1.4c. Similar trends are also found in ultra-low power radios with sub-GHz carrier from Fig. 1.4d and somewhat steeper with above-3 GHz carrier, without any conclusive trend in-between.

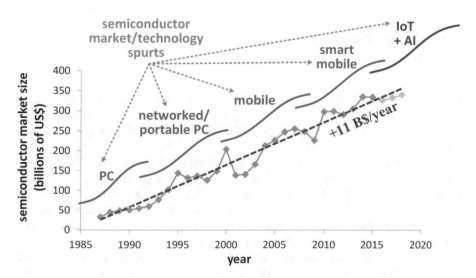

**Fig. 1.2**  Historical trend of computer size and global sale volume [1]

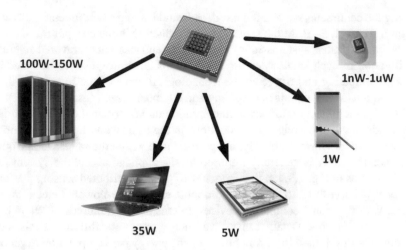

**Fig. 1.3** Power budget across different applications

Continuing the above energy down-scaling trends in the decade ahead is crucial, as the energy source form factor and cost are rapidly becoming dominant at the system level, and hence the bottleneck in the pursuit of further miniaturization and cost reduction. As widespread expectation, such energy reductions will have to be achieved while further improving performance, as needed by the natural evolution of recent and new applications.

In other words, both peak performance and energy efficiency need to be continuously increased to fulfill the demand for more capable computing systems, while sustaining the same miniaturization and cost reduction pace.

## 1.2 Energy Considerations in Power-Limited and Battery-Powered Systems

In the last decade, there has been a clear trend towards increased computation capabilities at the edge of distributed sensing/processing systems and hence in self-powered systems whose energy is provided by a local battery (or super-capacitor) and/or an energy harvester [7].

From the energy source viewpoint, Fig. 1.5 shows the battery life for different batteries versus the average power drawn by the system it powers. From this figure, a rechargeable thin-film battery with low 0.7 mm$^3$ volume can support sub-$\mu$W operation for only about a week, thus necessarily requiring the supplementary power coming from a harvester. The battery life increases to about a decade if the battery form factor is increased to button cell or larger size.

From a communication standpoint, typical short-range radios with energy per bit in the nJ/bit range and data rates of Mbps, the average power (i.e., the product of the

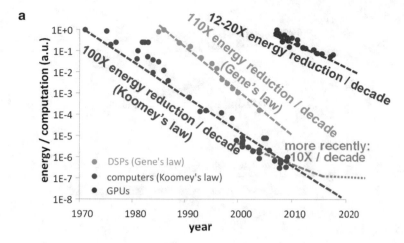

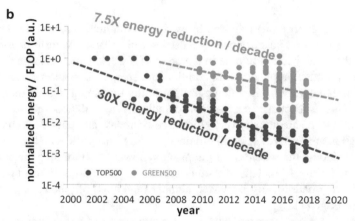

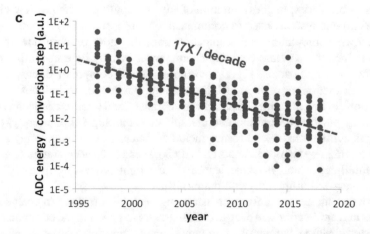

**Fig. 1.4** Energy trends [6] in (**a**) computers, digital signal processors, and GPUs, (**b**) supercomputers, (**c**) analog-to-digital converters, (**d**) short-range ultra-low power radios

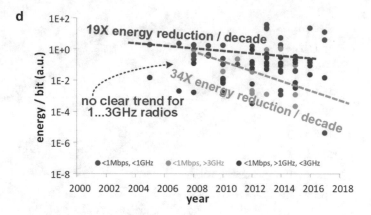

**Fig. .4** (continued)

data rate and the energy/bit) required by wireless communications is in the mW range. The latter cannot be sustained by batteries with reasonable form factor from Fig. 1.5, as the battery life would be in the order of a week, which is largely inadequate for any practical application. In addition, the energy/bit in short-range radios is expected to have very limited down-scaling in the upcoming decade (e.g., by a few units). This is because existing low-complexity modulation schemes (e.g., BPSK, BFSK, OOK) are already very close to best-in-class and power-hungry complex modulations in terms of spectral efficiency (e.g., M-ary FSK with large M) [1]. In addition, reducing the wireless communication power by shortening the communication range below a few tens of meters is not really an option in most applications, due to the quadratically increased number (and cost) of gateways gathering data from the edge [1].

The above-discussed large and hard-to-scale power cost of wireless communications explains the widespread exploitation of edge computing in the last few years, as more intelligent systems need to communicate more infrequently. According to the well-known computation-communication tradeoff, more on-chip intelligence permits to reduce the wireless communication power penalty by transmitting only aggregate data in small packets [1] (e.g., via compression, compressive sensing, on-chip feature extraction, or other signal dimensionality reduction techniques). As an additional option, few infrequent packets can be transmitted upon event occurrence, by embedding pattern recognition and classification on chip [1]. Currently, edge computing is evolving in several forms, including heterogeneous computing architectures (e.g., mixing microprocessors and GPUs), specialized accelerators (e.g., for Artificial Intelligence and machine learning), and tight processing-memory coupling (e.g., near- and in-memory computing) [8].

The above trend demands for ever-increasing peak performance to enable new capabilities and applications, at ever-decreasing average power in the common case. From a system-level viewpoint, the two opposite power-performance requirements have been usually decoupled through the introduction of duty cycling [9]. In this

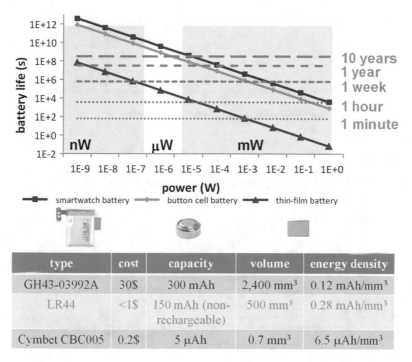

**Fig. 1.5** Battery life vs. average system power for different types of batteries [6]

case, the average system power is reduced by alternating active tasks and long sleep periods as depicted in Fig. 1.6, with periodicity set by the wake-up cycle $T_{wkup}$. Architecturally, an always-on sub-system manages the periodicity of the wake-up cycle and stores information across active tasks, drawing a constant power $P_{avg}$. At the same time, a duty-cycled sub-system periodically performs the active task at an energy $E_{duty\text{-}cycled}$. Accordingly, the overall average system power results to

$$P_{avg} = P_{always\ on} + \frac{E_{duty\ cycled}}{T_{wkup}}. \tag{1.1}$$

From Eq. (1.1), the average power can be reduced by reducing the power (energy) of the always-on (duty-cycled) sub-system and reducing the system activity by increasing $T_{wkup}$. Hence, nearly minimum energy per task needs to be pursued in the duty-cycled sub-system, to minimize the system power and prolong the battery life. As further consideration, Eq. (1.1) shows that duty cycling suffers from an inescapable compromise between low power consumption and infrequent wake-up, which translates into a higher probability to miss events that occur during operation in sleep mode. This limitation stems from the time-driven nature of duty-cycled systems, as signals cannot be monitored during sleep. From a performance standpoint, occasional occurrence of events detected during active tasks requires to be handled

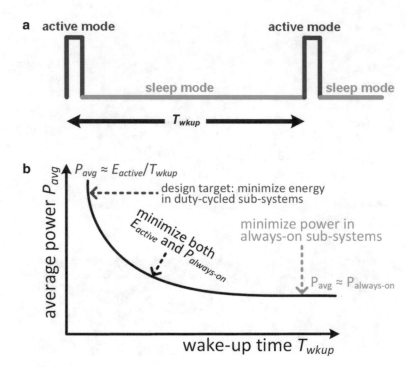

**Fig. 1.6** (**a**) Power mode sequence in duty-cycled systems, (**b**) qualitative trend plot of the average power vs. wake-up time in Eq. (1.1) [1]

with adequate responsiveness (i.e., latency), which in turn mandates adequately high performance when required.

In many recent and prominent applications, the above discussed tradeoffs are further tightened by the requirement that the probability to miss an event is nearly zero. This requires the system to be always on and to continuously monitor the inputs (e.g., coming from sensors) to detect events of interest in real time, to either transmit detailed data to the cloud or perform on-chip data analytics for further event characterization and understanding. In always-on systems, the monitoring task is performed at an energy $E_{\text{monitor}}$ and a specific rate $f_{\text{monitor}}$ that is set by the application (e.g., frame rate in vision tasks). When an event is detected, a more complex task (e.g., data analytics, selected data transmission to the cloud) is performed to gain the necessary knowledge from the input data or to take proper actions in response to the event. The resulting average power contribution $P_{\text{avg}}$ hence results to

$$P_{\text{avg}} = E_{\text{monitor}} \cdot f_{\text{monitor}} + \frac{E_{\text{task}}}{T_{\text{event,avg}}} \tag{1.2}$$

where $T_{\text{event, avg}}$ is the average time between two successive events. From Eq. (1.2), system power reductions require the minimization of the energy $E_{\text{monitor}}$ spent for each monitoring task, as the latter is performed continuously to avoid missing events. In addition, the average power is further reduced by reducing the energy $E_{task}$ associated with the occasional task performed upon event occurrence. Being the monitoring sub-system always-on, it can potentially capture any event and hence greatly relaxes the tradeoff between system power and the probability to capture all events. Indeed, assuming an ideal monitoring sub-system is available, no event is missed in event-driven systems. In practice, the probability to capture an event is high but not exactly 100%, as the accuracy of the monitoring sub-system to correctly detect events is never 100%. In other words, probability of correctly capturing events in time-driven systems is dictated by $T_{wkup}$, which in turn needs to be pessimistically set to the worst-case (shortest) time between two successive events. More quantitatively, this means that $T_{wkup}$ in Eq. (1.1) tends to be much smaller than $T_{event, avg}$ in Eq. (1.2), thus substantially increasing the power consumption associated with the task performed around events. Similar considerations hold in terms of response time after an event starts appearing for a long time. Instead, the activation rate of the data analysis task in event-driven systems is inherently adaptive and based on the actual event occurrence, thus avoiding the design pessimism of time-driven systems.

It is worth noting that the ability of event-driven systems to potentially capture all events is obtained at the cost of an always-on power $E_{monitor} \cdot f_{monitor}$ that generally exceeds the always-on power of duty-cycled systems. This justifies why nearly minimum energy needs to be pursued in always-on systems from Eq. (1.2), in terms of both $E_{monitor}$ and $E_{task}$. From a performance standpoint, occasional occurrence of events detected during active tasks requires to be handled with adequate responsiveness (i.e., latency), which again mandates adequately high performance when required.

## 1.3   Wide Power-Performance Tradeoff and System Requirements

### 1.3.1   Importance of Wide Power-Performance Tradeoff in Duty-Cycled and Always-On Systems

Several prominent applications are increasingly relying on a low energy requirement in the common case (i.e., low average power), while delivering much higher performance when occasionally needed. Although the performance requirement equally applies to both time- and event-driven systems, the energy requirement is even more crucial in the latter ones, due to the constant consumption of the monitoring sub-system in Eq. (1.2).

As an example of class of systems requiring wide power-performance tradeoff, next-generation intelligent sensor nodes for IoT are required to perform significant processing aboard to (1) reduce the communication power overhead (see previous section), (2) avoid the high cost of data deluge in the cloud infrastructure

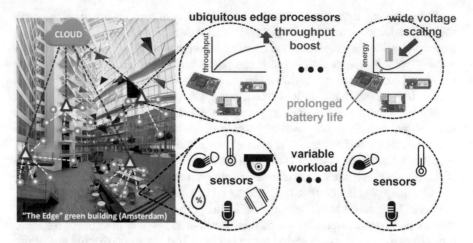

**Fig. 1.7** Intelligent sensor nodes with edge computing in a smart building. The figure shows the low-energy requirement in the common case and the need for rapid performance scale-up upon event occurrence

(e.g., storage, network bandwidth), (3) mitigate security and privacy issues, and (4) improve the overall system efficiency and responsiveness by feeding the cloud with relevant data, rather than raw data. This class of applications is exemplified by smart building control as in Fig. 1.7, where a plethora of sensor nodes are distributed around the environment and embedded in objects to monitor environmental conditions, resource utilization, occupancy, occupants' activity, and so forth. To make the building able to meet to the occupants' needs, responsive actions are required to be in the scale of very few seconds in terms of lighting, air conditioning and ventilation (including pre-emptive activation to prepare rooms for specific purposes), access control, access to media resources, and so forth. Edge devices need to capture such events expeditiously and occasionally need to substantially scale up their performance to gain deeper knowledge of the context before sending data to the cloud. For example, such deeper knowledge of occurring events might require the execution of machine learning for refined event detection, or to increase the number of sensors that are simultaneously monitored (e.g., sensor fusion) to extract valuable and actionable knowledge. On the other hand, the change in such patterns is much slower and can be easily in the scale of several minutes or tens of minutes. Hence, events detected at the edge are rather infrequent, thus vastly limiting the occurrence of on-chip complex computation.

As another example of systems requiring wide power-performance tradeoff, always-on self-powered speech recognition systems draw an unacceptably large power levels if kept on all the time. Accordingly, they are usually equipped with a front-end sub-system performing voice activity detection (VAD), i.e., spotting whether speech events are occurring. In the common case where no speech is taking place, the speech recognition engine is clock gated to keep its power negligible (see Fig. 1.8). Upon speech event occurrence, the engine operates to recognize the

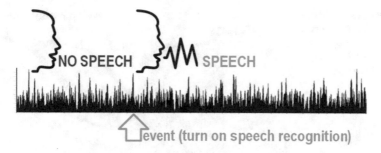

**Fig. 1.8** Event-driven speech recognition through always-onvoice activity detection

content of the speech and returns to the sleep mode thereafter. Similar considerations can be made for systems for audio context awareness (e.g., surveillance systems) where salient features (speeding cars, unauthorized entrance, etc.) have to be detected and responded to.

A wide range of other applications require wide power-performance tradeoff and event-driven operation, which have become a major challenge and key enabler of new applications in computer vision [10–12], audio monitoring [13–15], sensing [16], energy metering [17], intelligent wireless communications [18, 19], among the many others. Such wide power-performance tradeoff in the digital sub-system is traditionally achieved through wide voltage scaling, as discussed in the following subsection.

### 1.3.2 Wide Voltage Scaling

The dynamic nature of the workload in systems with wide power-performance tradeoff has motivated the adoption of wide dynamic voltage frequency scaling (DVFS) [20–27]. DVFS allows the digital sub-system to run at the lowest required performance target to save energy via aggressive voltage scaling, whereas peak performance is achieved by operating at the nominal voltage allowed by the process [20].

The energy-performance tradeoff under wide DVFS is qualitatively shown in Fig. 1.9. Analytically, the total energy per cycle $E_{cycle}$ in the digital sub-system is equal to the sum of the dynamic energy ($E_{dyn}$) and the leakage energy ($E_{lkg}$):

$$E_{cycle} = E_{dyn} + E_{lkg} = \alpha_{sw} \cdot C_{tot} \cdot V_{DD}^2 + V_{DD} \cdot I_{off} \cdot T_{CK}, \tag{1.3}$$

where $C_{tot}$ is the total physical capacitance associated with the logic gates and the interconnects, $\alpha_{sw}$ is the switching activity factor (i.e., the fraction of $C_{tot}$ that is switched every cycle on average), $V_{DD}$ is the supply voltage, $I_{off}$ is the average overall leakage current, and $T_{CK}$ is the clock period [28].

From Eq. (1.3), the dynamic energy quadratically decreases when $V_{DD}$ is scaled down. On the other hand, the leakage energy increases nearly exponentially in the

**Fig. 1.9** Qualitative trend
of gate delay (red curve)
and energy per operation
(blue curve) vs. supply
voltage $V_{DD}$ [1, 9]

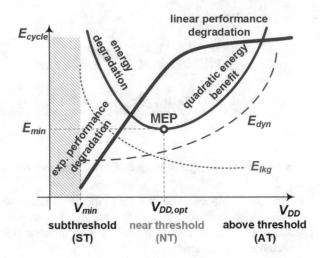

sub-threshold region, due the exponential increase in the gate delay (and hence $T_{CK}$) [1, 9]. From Fig. 1.9, a minimum-energy point (MEP) is observed at the sub- or the near-threshold voltage region that optimally balances the dynamic and leakage energy, thus leading to the minimum energy $E_{\min}$. From a performance perspective, occasional speed boosts are achieved by raising $V_{DD}$ from the MEP to the nominal voltage.

Wide voltage scaling can reduce the energy by up to an order of magnitude, when comparing the MEP to operation at nominal voltage [1]. The performance can be scaled up by one to two orders of magnitude when scaling $V_{DD}$ from the MEP to nominal voltage [1, 20–27]. Such performance increase is clearly achieved at the expense of the above-mentioned increase in the energy per operation. From this figure, this energy increase is more pronounced when the MEP is in the sub-threshold region, due to the larger voltage difference between the MEP and the nominal voltage.

## 1.4  Challenges in Wide Voltage Scaling and Motivation

Wide voltage scaling poses a fundamental dilemma to the designer, as optimizing the design at the one end of the power-performance spectrum inevitably degrades the other end. This holds both for designs optimized at a fixed voltage and then extended to other targets through DVFS and for designs using multi-mode multi-corner methodologies where the contradicting constraints at different voltages are managed to find a common design optimization that meets all of them. In any of those two design scenarios, statically defined design parameters cannot be adapted to the different needs arising at different voltages.

From a design standpoint, the adoption of a wide voltage range introduces a wide shift in two fundamental tradeoffs, which makes the overall digital circuit

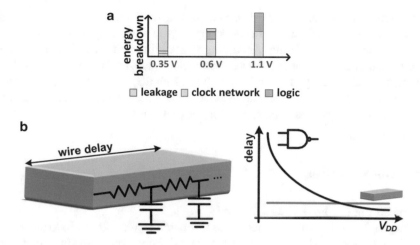

**Fig. 1.10** (**a**) Energy breakdown vs. $V_{DD}$ as measured in an FFT testchip (see Chap. 4), (**b**) qualitative trend of wire and gate delay vs. $V_{DD}$ [1]

optimization very intricate. As first voltage-sensitive tradeoff, the composition of the overall energy of the digital sub-system drastically changes when moving from above-threshold down to near- and sub-threshold voltages [1, 29–33]. As an example, and as extensively discussed in Chap. 2, Fig. 1.10a shows the energy breakdown of a testchip performing the Fast Fourier Transform (FFT) from sub-threshold to above-threshold voltages. From this figure, the clocking and the combinational dynamic energy contributions $E_{dyn}$ dominate, whereas the leakage energy $E_{lkg}$ in Eq. (1.3) is negligible at the nominal 1.1 V supply voltage. When the latter is scaled down to near-threshold voltages around 0.6 V, the leakage energy starts being important. At sub-threshold voltages around 0.35 V, the leakage energy definitely dominates over the other contributions. At the same time, the leakage–dynamic energy ratio $E_{lkg}/E_{dyn}$ is well known to fundamentally determine the energy-optimal microarchitecture. Hence, a traditional fixed microarchitecture cannot be optimized to achieve its highest energy efficiency, as $E_{lkg}/E_{dyn}$ varies substantially under wide DVFS from Fig. 1.10a.

As a second fundamental and voltage-sensitive tradeoff, the wire delay scales very differently compared with the gate delay, as summarized in Fig. 1.10b. Indeed, the wire delay is solely determined by its RC time constant, which in turn does not change when scaling the supply voltage. On the other hand, the gate delay linearly increases when scaling down $V_{DD}$ in the above-threshold region, whereas it increases super-linearly in the near-threshold region, and then exponentially in the sub-threshold region. In turn, the ratio of the wire and gate delay is well known to determine the correct topology of the clock distribution network that achieves a reasonable tradeoff between energy, performance, and clock skew (see Chap. 5). Accordingly, a traditional fixed clock network cannot be optimized to have a nearly optimal balance between wire and gate delay at all supply voltages [34, 35].

In summary, optimizing the data path with a fixed microarchitecture and the clock path with a fixed clock network topology is invariably sub-optimal when adopting wide voltage scaling. Accordingly, flexibility and reconfiguration ability need to be introduced in both the data and the clock path, as discussed in the following chapters. Ultimately, data and clock path adaptation allow to outperform traditional voltage scaling, providing an extended power-performance range as required in the prominent applications discussed in the previous section.

## 1.5   Book Outline

**The outline of the book is as follows:**
- Chapter 2 of this book introduces the microarchitectural knobs in logic and on-chip memories that enable the extension of the energy-performance tradeoff beyond voltage scaling.
- Chapter 3 describes the design methodologies required to integrate these knobs into an automated design flow, enabling data path reconfiguration at minimal design effort.
- Chapter 4 discusses several case studies and testchip measurements to quantify the advantages and the penalties brought by reconfigurable microarchitectures.
- Chapter 5 focuses on the clock path reconfiguration, the related design methodologies and flow, and a case study.
- Chapter 6 concludes the book and summarizes the findings in the previous chapters.
- Finally, the Appendix describes a complete set of scripts that allow the immediate implementation of the design flows presented in the previous chapters. To make this book useful and immediately applicable to a wide range of applications, research prototypes, and IC products, the scripts have been made publicly available through [36].

## References

1. M. Alioto (ed.), *Enabling the Internet of Things—From Integrated Circuits to Integrated Systems* (Springer, Berlin, 2017)
2. W. Rhines, Gompertz predicts the future. (SemiWiki, 2019 [Online]), https://semiwiki.com/wally-rhines/273854-chapter-four-gompertz-predicts-the-future/. Accessed 2 Aug 2019
3. C.G. Bell, R. Chen, S. Rege, Effect of technology on near term computer structures. IEEE Comput. **5**(2), 29–38 (1972)
4. G. Bell, Bell's law for rise and death of computer classes. Commun. ACM **51**(1), 86–94 (2008)
5. M. Fojtik, D. Kim, G. Chen, Y.-S. Lin, D. Fick, J. Park, M. Seok, M.-T. Chen, Z. Foo, D. Blaauw, D. Sylvester, A millimeter-scale energy-autonomous sensor system with stacked battery and solar cells. IEEE J. Solid State Circuits **48**(3), 801–813 (2013)

6. M. Alioto, V. De, A. Marongiu, Energy-quality scalable integrated circuits and systems: continuing energy scaling in the twilight of Moore's law. IEEE J. Emerg. Select. Topics Circuits Syst. **8**(4), 653–678 (2018)

7. M. Alioto, Energy harvesters for IoT: applications and key aspects—short course at *VLSI Symposium 2015*, Kyoto (Japan). Accessed 15 June 2015

8. ISSCC Trends [Online]. http://isscc.org/isscc-in-the-news-2/trends/

9. M. Alioto, Ultra-low power VLSI circuit design demystified and explained: a tutorial. IEEE Trans. Circuits Syst. Pt. I (Invited) **59**(1), 3–29 (2012)

10. L. Camunas-Mesa, C. Zamarreno-Ramos, A. Linares-Barranco, A.J. Acosta-Jimenez, T. Serrano-Gotarredona, B. Linares-Barranco, An event-driven multi-kernel convolution processor module for event-driven vision sensors. IEEE J. Solid State Circuits **47**(2), 504–517 (2012)

11. M. Rusci, D. Rossi, M. Lecca, M. Gottardi, E. Farella, L. Benini, An event-driven ultra-low-power smart visual sensor. IEEE Sens. J. **16**(13), 5344–5353 (2016)

12. G. Singh, A. Nelson, S. Lu, R. Robucci, C. Patel, N. Banerjee, Event-driven low-power gesture recognition using differential capacitance. IEEE Sens. J. **16**(12), 4955–4967 (2016)

13. M. Price, J. Glass, A.P. Chandrakasan, A scalable speech recognizer with deep-neural-network acoustic models and voice-activated power gating, in *IEEE International Solid-State Circuits Conference (ISSCC) Digest of Technical Papers*, (2017), pp. 244–245

14. M. Yang, C. Chien, T. Delbruck, S. Liu, A 0.5 V 55μW 64x2-channel binaural silicon cochlea for event-driven stereo-audio sensing. IEEE J. Solid State Circuits **51**(11), 2554–2569 (2016)

15. S. Lecoq, J. Le Bellego, A. Gonzalez, B. Larras, A. Frappé, Low-complexity feature extraction unit for "Wake-on-Feature" speech processing, in *Proceedings of ICECS*, (2018), pp. 677–680

16. C. Weltin-Wu, Y. Tsividis, An event-driven clockless level-crossing ADC with signal-dependent adaptive resolution. IEEE J. Solid State Circuits **48**(9), 2180–2190 (2013)

17. M. Simonov, G. Chicco, G. Zanetto, Event-driven energy metering: principles and applications. IEEE Trans. Indus. Appl. **53**(4), 3217–3227 (2017)

18. Y. Xiao, W. Li, M. Siekkinen, P. Savolainen, A. Ylä-Jääski, P. Hui, Power management for wireless data transmission using complex event processing. IEEE Trans. Comput. **61**(12), 1765–1777 (2012)

19. X. Huang, P. Harpe, G. Dolmans, H. de Groot, J.R. Long, A 780–950 MHz, 64–146 μW power-scalable synchronized-switching OOK receiver for wireless event-driven applications. IEEE J. Solid State Circuits **49**(5), 1135–1147 (2014)

20. A. Chandrakasan, D. Daly, D. Finchelstein, J. Kwong, Y. Ramadass, M. Sinangil, V. Sze, N. Verma, Technologies for ultradynamic voltage scaling. Proc. IEEE **98**(2), 191–214 (2010)

21. H. Kaul, M.A. Anders, S.K. Mathew, S.K. Hsu, A. Agarwal, F. Sheikh, R.K. Krishnamurthy, S. Borkar, A 1.45GHz 52-to-162GFLOPS/W variable-precision floating-point fused multiply-add unit with certainty tracking in 32nm CMOS, in *IEEE ISSCC Digest of Technical Papers*, (2012), pp. 182–183

22. J. Myers, A. Savanth, D. Howard, R. Gaddh, P. Prabhat, D. Flynn, An 80nW retention 11.7pJ/cycle active sub-threshold ARM Cortex®-M0+ Sub-System in 65nm CMOS for WSN applications, in *IEEE ISSCC Digest of Technical Papers*, (2015), pp. 144–145

23. S. Hsu, A. Agarwal, M. Anders, S. Mathew, H. Kaul, F. Sheikh, R. Krishnamurthy, A 280mV-to-1.1V 256b reconfigurable SIMD vector permutation engine with 2-dimensional shuffle in 22nm CMOS, in *ISSCC Digest of Technical Papers (ISSCC), San Francisco (CA)*, (2012)

24. J. Wang, N. Pinckney, D. Blaauw, D. Sylvester, Reconfigurable self-timed regenerators for wide-range voltage scaled interconnect. Proc. ASSCC **2015**, 18–15 (2015)

25. S. Jain et al., A 280mV-to-1.2V wide-operating-range IA-32 processor in 32nm CMOS, in *IEEE ISSCC Digest of Technical Papers*, (2012), pp. 66–67

26. F. Sheikh, S. Mathew, M. Anders, H. Kaul, S. Hsu, A. Agarwal, R. Krishnamurthy, S. Borkar, A 2.05GVertices/s 151mW lighting accelerator for 3D graphics vertex and pixel shading in 32nm CMOS, in *IEEE ISSCC Digest of Technical Papers*, (2012), pp. 178–179

27. G. Gammie, N. Ickes, M. Sinangil, R. Rithe, J. Gu, A. Wang, H. Mair, S. Datla, R. Bing, S. Honnavara-Prasad, L. Ho, G. Baldwin, D. Buss, A. Chandrakasan, U. Ko, A 28nm 0.6V low-power DSP for mobile applications, in *ISSCC Digest of Technical Papers (ISSCC), San Francisco (CA)*, (2011)
28. S. Jain, L. Lin, M. Alioto, Design-oriented energy models for wide voltage scaling down to the minimum energy point. IEEE Trans. CAS Pt. I (TCAS-I) **64**(12), 3115–3125 (2017)
29. S. Jain, L. Lin, M. Alioto, Dynamically adaptable pipeline for energy-efficient microarchitectures under wide voltage scaling. IEEE J. Solid-State Circuits **53**(2), 632–641 (2018)
30. S. Jain, L. Lin, M. Alioto, Drop-in energy-performance range extension in microcontrollers beyond VDD scaling, in *IEEE Asian Solid-State Circuits Conference (A-SSCC)*, (2019)
31. S. Jain, L. Lin, M. Alioto, Automated design of reconfigurable micro-architectures for accelerators under wide voltage scaling, in *IEEE Transactions on Very Large Scale Integration (VLSI) Systems*, (2019), pp. 1–14
32. L. Lin, S. Jain, M. Alioto, A 595pW 14pJ/cycle microcontroller with dual-mode standard cells and self-startup for battery-indifferent distributed sensing, in *IEEE ISSCC Digest of Technical Papers*, (2018), pp. 44–45
33. L. Lin, S. Jain, M. Alioto, Integrated power management and microcontroller for ultra-wide power adaptation down to nW, in *VLSI Symposium 2019, Kyoto (Japan)*, (2019), pp. C178–C179
34. L. Lin, S. Jain, M. Alioto, Reconfigurable clock networks for wide voltage scaling. IEEE J. Solid State Circuits **54**(9), 2622–2631 (2019)
35. L. Lin, S. Jain, M. Alioto, Reconfigurable clock networks for random skew mitigation from subthreshold to nominal voltage, in *IEEE ISSCC Digest of Technical Papers, San Francisco (CA)*, (2017), pp. 440–441
36. S. Jain, M. Alioto, RECMICRO: design framework and scripts to design reconfigurable micro-architectures [Online], http://www.green-ic.org/recmicro.

# Chapter 2
# Reconfigurable Microarchitecures Down to Pipestage and Memory Bank Level

**Abstract** As discussed in Chap. 1, no single fixed microarchitecture can be energy-optimal over a wide range of supply voltages. Accordingly, microarchitecture reconfiguration is needed to enhance the energy efficiency over the entire voltage range and to expand the power-performance tradeoff beyond allowed by voltage scaling. Similar considerations hold for the memory, as it needs to be reconfigured down to the bank level to consistently offer access time improvements when the logic depth (i.e., the clock cycle) of the microarchitecture is modified through reconfiguration.

**Keywords** Reconfigurable microarchitecture · Reconfigurable SRAM · Pipestage-level reconfiguration · Thread-level reconfiguration · Bank-level reconfiguration · Wide voltage scaling · Energy consumption · Power-performance tradeoff extension · Synchronous digital systems · Timing constraints · Flip-flop · Pipestage · Pipeline stage · Launching register · Capturing register · Setup time · Hold time · Clock-to-Q delay · Positive edge-triggered · Negative edge-triggered · Dual edge-triggered · Clock edge · Timing overhead · Clock distribution · Effective logic depth · Fan-out-of-4 delay · Pipelining · Re-pipelining · Throughput · Latency · Area · Balanced pipestages · Dataflow · Delay imbalance · Pipeline depth · Pipedepth · FIR filters · Fast Fourier Transform (FFT) · IIR filters · Register level · Linear pipeline · Feedforward pipeline · Feedback pipeline · Loops · Dynamic energy · Leakage energy · Activity factor · Switched capacitance · Clock cycle · Leakage current · Above-threshold region · Near-threshold region · Sub-threshold region · Leakage-dynamic energy ratio · Fixed microarchitectures · minimum energy point (MEP) · Shallow pipelines · Deep pipelines · Energy breakdown · Dynamically adaptable pipelines · Dynamic voltage frequency scaling · DVFS · Wide DVFS · Clock generator · Phase-locked loop · Clock frequency · DSP · Voltage regulator · Power mode · Bypassable register · Bypassable flip-flop · Non-bypassable register · Non-bypassable flip-flop · EDA tool · Retiming · Delay overhead · Look-up table · Register bypassing · Flip-flop bypassing · Cross-point voltage · Microprocessor · Thread · Always-on systems · Throughput enhancement · Control flow · Hazards · Pipeline bubble · Microprocessor microarchitecture · Microarchitecture · RISC processor · Instruction execution · Pipeline stall · Stall · Compiler · Time-interleaved microarchitecture · Time interleaving · Input stream · Instruction stream · Channel · Processing element · Dynamically adaptable time-interleaved microprocessor · Logic depth · Logic depth adjustment · Logic

© Springer Nature Switzerland AG 2020
S. Jain et al., *Adaptive Digital Circuits for Power-Performance Range beyond Wide Voltage Scaling*, https://doi.org/10.1007/978-3-030-38796-9_2

gate · Gate-level netlist · Software stack · Logic depth-voltage co-optimization ·
Microarchitecture-voltage co-optimization · General-purpose processor ·
Application-specific design · SRAM · Instruction fetch · Instruction decode ·
Execute · Instruction memory · Data memory · Column multiplexing · Column
multiplexer · Bitline · Wordline · Sense amplifier · Precharge driver · Write driver ·
Bitcell · Memory bank · Memory sub-bank · Floorplan · Read current · Bitline
capacitance · Row · Column · Row decoder · Column decoder · Reconfigurable
array organization · Access time · Row aggregation · Pulsewidth · Bitline discharge
· Memory address space

As discussed in Chap. 1, no single fixed microarchitecture can be energy-optimal
over a wide range of supply voltages. Accordingly, microarchitecture reconfiguration is needed to enhance the energy efficiency over the entire voltage range and to
expand the power-performance tradeoff beyond allowed by voltage scaling. Similar
considerations hold for the memory, as it needs to be reconfigured down to the bank
level to consistently offer access time improvements when the logic depth (i.e., the
clock cycle) of the microarchitecture is modified through reconfiguration.

This chapter introduces dynamically adaptable microarchitectures and SRAM
memories enabling joint microarchitectural and voltage co-optimization at run-
time, from nominal voltage down to deep sub-threshold. The additional flexibility
offered by such dynamic adaptation offers an extra knob that adds to conventional
voltage scaling, extending the power-performance tradeoff. Reconfiguration techniques are introduced at the pipeline level in accelerators, thread level in microprocessors, and bank level in SRAMs.

## 2.1   Pipestage as Basic Building Block of Synchronous Microarchitectures

### 2.1.1   Background on Pipeline Stages and Timing Constraints

The fundamental building block of any synchronous microarchitecture is a pipeline
stage, also called "pipestage" [1, 2]. A pipestage consists of a launching register to
synchronize the input to the clock CLK, an intermediate combinational logic performing the pipestage computation, and a capturing register that synchronizes its output to the clock (see Fig. 2.1a). The launching and capturing registers are generally
flip-flops as assumed in the rest of the book, although they might sometimes be of a
different nature [1] (e.g., latches, pulsed latches). Positive (negative) edge-triggered
flip-flops synchronize their input signal to the clock around its rising (falling) edge,
whereas double edge-triggered flip-flops synchronize around all edges [2].

As recalled in Fig. 2.1b, flip-flops impose timing constraints on when the input
can transition with respect to the active edge of the clock, which is assumed to be

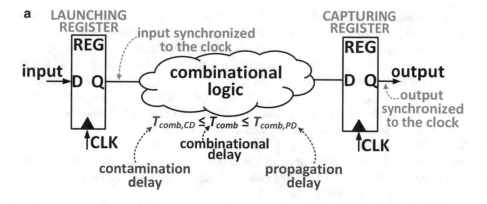

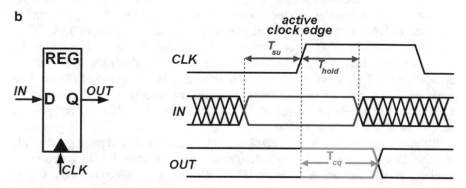

**Fig. 2.1** (a) Structure of a generic pipestage, (b) timing constraints and parameters in flip-flop-based registers (positive edge-triggered in this example)

rising in Fig. 2.1b with no loss of generality. Indeed, the input needs to be stabilized at a setup time $T_{su}$ before the clock edge, and hence needs to be available early enough compared to the next clock edge. At the same time, the input needs to start transitioning again only a hold time $T_{hold}$ after the clock edge and hence needs to change late enough compared to the current edge. These two flip-flop timing constraints respectively translate into an upper and a lower bound of the allowed pipestage combinational delay [1, 2].

Regarding the setup time constraint, the combinational output needs to be available $T_{su}$ before the next clock edge at the capturing register. From Fig. 2.1, this means that $T_{CK}$ needs to be long enough to include the delay for the generation of the launching register output (i.e., flip-flop clock-to-Q delay $T_{cq}$), the worst-case delay of the combinational logic (i.e., propagation delay $T_{comb, PD}$ in Fig. 2.1a), the setup time of the capturing flip-flop $T_{su}$, and the additional clock uncertainty $T_{cu}$ due to the non-idealities in the clock generation and distribution [1] (as set by the difference in the arrival time of the clock at the capturing over the launching register):

$$T_{CK} \geq T_{cq} + T_{comb,PD} + T_{su} + T_{cu} = T_{comb,PD} + \tau_{OH}. \tag{2.1}$$

From Eq. (2.1), only part of the clock cycle is actually available for useful computation in the combinational logic (i.e., $T_{comb, PD}$), whereas the remaining part of the clock cycle $\tau_{OH} = T_{CK} - T_{comb, PD}$ represents a timing overhead that needs to be born due to synchronous operation. The contribution of flip-flops to $\tau_{OH}$ is $T_{cq} + T_{su}$ and is due to the timing penalty of the launching and the capturing register. The remaining contribution to $\tau_{OH}$ is due to (1) the temporal fluctuations of $T_{CK}$ due to the clock generation and distribution (i.e., jitter [1]); (2) the clock skew between the launching and the capturing register due to their clock arrival time misalignment (as determined by the non-uniform delay through the clock distribution network). The sign of $T_{cu}$ can be either positive or negative, depending on how the clock distribution direction compared to the data flow direction and on how large the random components of the above contributions are [1].

From Fig. 2.1, the clock frequency $f_{CK} = 1/T_{CK}$ increases when reducing the combinational propagation delay, which in turn can be achieved by reducing the number of cascaded logic gates in the critical logic path (named "logic depth") within the pipestage. From Eq. (2.1), logic depth reductions translate into significantly better performance as long as $T_{comb, PD} \gg \tau_{OH}$, whereas rapidly diminishing returns are achieved when $T_{comb, PD}$ becomes only a few times $\tau_{OH}$. Hence, practical values of the logic depth are both upper and lower bounded, as excessive values prevent the achievement of the targeted $f_{CK}$, whereas excessively low values do not provide much benefit and increase the clocking energy.[1]

To gain a better sense of the magnitude of the clock cycle and the logic depth in a technology-independent manner, designers usually refer to the clock cycle normalized to the fan-out-of-4 delay (FO4) [2] (named "effective logic depth" $LD_{eff} = T_{CK}/FO4$, expressed as number of FO4/pipestage). FO4 is defined as the delay of an inverter gate driving four equal inverter gates, and approximates well the delay of any logic gate resulting from gate-level optimization for maximum speed [1–3]. For slower design targets, the gate delay becomes larger than FO4. Hence, the ratio $T_{CK}/FO4$ represents the effective number of gate delays in a cycle (or equivalently in a pipestage) in high-speed designs and an upper bound of the logic depth in slower designs. As the gate delay normalized to FO4 is nearly independent of process, voltage, and temperature, $LD_{eff}$ is solely defined by the microarchitecture [3]. Generally speaking, the above-mentioned $f_{CK}$ speed-ups are achieved when the effective logic depth is in the order of a few tens of FO4 per stage, or more. Instead, diminishing returns are obtained under deep microarchitectures with logic depth of roughly 20 FO4 per stage or lower, and marginal gains are achieved when approaching the 10 FO4/pipestage range [2].

Regarding the hold time constraint, from Fig. 2.1b the combinational logic output is allowed to start transitioning only a hold time $T_{hold}$ after the clock edge. This translates into a lower bound to the minimum combinational delay (i.e., contamina-

---

[1] This is because a substantially larger number of registers needs to be inserted to reduce the logic depth in each pipestage, leading to a disproportionate number of flip-flops being driven by the clock distribution network.

tion delay $T_{comb, CD}$ in Fig. 2.1a), since hold violations simply occur when the latter reacts too fast to a new input [1, 2]. In particular, the condition that suppresses hold violations is well known to be [1–3]

$$T_{comb,CD} \geq T_{hold} - T_{cq} + T_{cu}.$$ (2.2)

From Eqs. (2.1) and (2.2), the clock uncertainty needs to be kept small enough to limit the performance degradation due to the additional timing overhead $T_{cu}$ in Eq. (2.1) and to suppress the occurrence of hold violations for a given combinational block from Eq. (2.2). More details on the clock uncertainty, its components, and experimental characterization will be provided in Chap. 5, where clock distribution networks will be covered.

## 2.1.2 Pipelining for Microarchitecture Speed-Up

From the previous subsection, reducing the combinational logic delay immediately translates into increased clock frequency $f_{CK}$ from Eq. (2.1) and hence improved throughput (i.e., number of computations per second). In detail, the combinational delay is effectively reduced by breaking down each combinational block in Fig. 2.1a into multiple shorter ones. In turn, this is achieved by inserting additional registers within the original combinational logic, changing the single pipestage in Fig. 2.1a to multiple pipestages in Fig. 2.2. The resulting speed-up comes at the cost of increased clocking energy as discussed in Sect. 2.3.

Quantitatively, the clock frequency improvement when a single pipestage is broken into multiple pipestages is given by

$$\frac{f_{CK,pipe}}{f_{CK}} = \frac{T_{CK}}{T_{CK,pipe}} = n \cdot \frac{T_{comb} + \tau_{OH}}{T_{comb} + n \cdot \tau_{OH}}$$ (2.3)

where $f_{CK, pipe}$ is the maximum clock frequency after splitting the pipestage into $n$ perfectly balanced pipestages (i.e., each with equal delay $T_{comb}/n$) as in Fig. 2.2, and

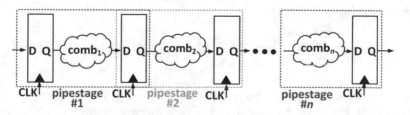

**Fig. 2.2** Example of single pipestage in Fig. 2.1a broken up into $n$ pipestages through re-pipelining (i.e., insertion of intermediate registers) with balanced combinational delays

$f_{CK}$ is the clock frequency of the original single pipestage with overall combinational delay $T_{comb}$ in Fig. 2.1a. The clock frequency speed-up in Eq. (2.3) translates into an actual throughput improvement by the same factor only if the throughput is not affected by re-pipelining. This might not be true when re-pipelining creates additional data interdependency between pipestages (e.g., in microprocessors), which would otherwise require previous computations to be occasionally stalled while waiting for the subsequent ones, as discussed at the end of this subsection. Under the above assumption, from Eq. (2.3) the throughput gain coming from re-pipelining is essentially equal to $n$ in the typical case when the latter is moderately high (i.e., such that $\tau_{OH} \ll T_{comb}/n$). Under aggressive re-pipelining, $n$ becomes so large that the combinational delay $T_{comb}/n$ after re-pipelining becomes comparable to the timing overhead $\tau_{OH}$, and rapidly diminishing returns are achieved for higher $n$.

In applications requiring system responsiveness, throughput is not a critical target and performance is better quantified by the latency, which is defined as the time required to deliver the output corresponding to a given input, once the input has become available. Due to the presence of the intermediate registers, the latency of the re-pipelined design inevitably increases. Quantitatively, from Eq. (2.1) the latency degradation of the re-pipelined design compared to the original one is readily evaluated to be

$$\frac{LAT_{pipe}}{LAT} = \frac{n \cdot T_{CK,pipe}}{T_{CK}} = n \frac{\dfrac{T_{comb,PD}}{n} + \tau_{OH}}{T_{comb,PD} + \tau_{OH}} \approx n \frac{\dfrac{\tau_{OH}}{T_{comb,PD}}}{1 + \dfrac{\tau_{OH}}{T_{comb,PD}}} \approx 1 + (n-1) \cdot \frac{\tau_{OH}}{T_{comb,PD}} \quad (2.4)$$

where $LAT_{pipe}$ is the latency of the re-pipelined design in Fig. 2.2 (i.e., equal to $n$ cycles, although each cycle is shorter than the original one), and $LAT$ is the latency of the original design in Fig. 2.1a. In Eq. (2.4), $n \ll T_{comb, PD}/\tau_{OH}$ was assumed as discussed above, and the result was approximated by the Taylor series truncated to the first order. From Eq. (2.4), the latency degradation due to re-pipelining is linear, and is quite slower than the clock frequency improvement, since in Eq. (2.4) the term $n$ is multiplied by term $\tau_{OH}/T_{comb, PD}$ that is much lower than one.

Similarly, the inevitable area increase due to the additional intermediate registers under re-pipelining is given by

$$\frac{A_{pipe}}{A} = \frac{A_{comb} + (n+1)A_{reg}}{A_{comb} + 2A_{reg}} \approx 1 + (n-1) \cdot \frac{A_{reg}}{A_{comb}} \quad (2.5)$$

where $A_{pipe}$ is the overall area after re-pipelining in Fig. 2.2, $A$ is the area of the original design in Fig. 2.1a, $A_{reg}$ is the area of each pipeline register, and $A_{comb}$ is the overall area of the combinational logic in both designs. Again, from Eq. (2.5) the area penalty of re-pipelining is linear with $n$ and increases slowly with it since the latter is multiplied by $A_{reg}/A_{comb} \ll 1$. Similarly, the energy overhead of the additional registers at the same supply voltage is equal to

$$\frac{E_{pipe}}{E} = \frac{E_{comb} + (n+1)E_{reg}}{E_{comb} + 2E_{reg}} \approx 1 + (n-1) \cdot \frac{E_{reg}}{E_{comb}} \tag{2.6}$$

where $E_{pipe}$ ($E$) is the energy after (before) re-pipelining, $E_{reg}$ is the average energy consumption of each pipeline register, and $E_{comb}$ is the overall energy of the combinational logic in both designs.

Figure 2.3 shows the dataflow in a pipelined microarchitecture. After the above-discussed $n$-cycle latency, the pipeline is completely filled as each pipestage has been reached by meaningful data and runs in parallel with all others. As in Eq. (2.3), a throughput improvement by a factor of $n$ is achieved if re-pipelining does not introduce any additional data inter-dependency. This is generally true when the microarchitecture can be properly crafted to suppress (or predictively deal with) data inter-dependency among the computations executed simultaneously in the different pipestages. This is typically the case for application-specific designs, as the data inter-dependencies have predictable patterns in view of the specificity of the microarchitecture. Instead, such patterns are generally not predictable in general-purpose systems (e.g., processors), as they need to assure full flexibility to execute any software code. In this case, the control flow needs to be enhanced to deal with such hazards and to stall computations if the related inputs are not yet available from the previous computations within the same pipeline. The stalls inevitable

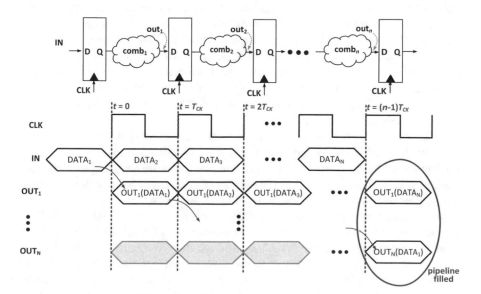

**Fig. 2.3** Dataflow in pipelined microarchitectures

cause a throughput degradation that is usually modeled by the average stall cycles per cycle $\delta$, which is defined as the average number of cycles that a computation needs to be stalled for each cycle of computation [4]. Accordingly, the throughput of the re-pipelined design accounting for stalls is degraded by a factor $(1 + \delta)$.

The above considerations were made for the generic pipestage in the original design before re-pipelining. The minimum clock cycle under an arbitrary number of original pipestages is immediately derived by considering that $T_{CK}$ needs to accommodate the delays in the slowest pipestage, and hence take the maximum clock cycle in Eq. (2.1) across all pipestages. Also, the above derivations are based on the assumption that all pipestages are well balanced. In practical cases, the delays across pipestages are not exactly the same, and the resulting delay imbalance results in slower operation. Simple calculations show that the effect of delay imbalance is equivalent to an additional overhead term in Eq. (2.1) that is given by the maximum pipestage delay deviation compared to the average, across all pipestages.

## 2.2 Elementary Microarchitectures

In general, any microarchitecture can be broken into portions that belong to three basic elementary microarchitectures, namely linear, feedforward, and feedback pipelines.

A linear pipeline consists of a number of *pipedepth* cascaded pipestages (*pipedepth* being the pipeline depth, i.e., the number of pipestages encountered from inputs to outputs). In linear pipelines, there is no pipestage branching at any pipestage, as depicted in Fig. 2.4a. Arithmetic multipliers are examples of linear pipelines. By definition of linear pipeline, any input-to-output path crosses the same number *pipedepth* of registers. Accordingly, the output at the generic $n$th cycle depends only on the input at the $(n - pipedepth)$-th cycle. In other words, the number of registers crossed in any path going from the inputs to a given intermediate register (named "register level") is independent of the specific gate-level path, as thoroughly discussed in Chap. 3.

In feedforward pipelines, there is at least one feedforward path that skips one or more levels of registers, as shown in Fig. 2.4b. As depicted in this figure, feedforward pipelines have pipestage-level branches. FIR filters are an example of feedforward pipelines. In such pipelines, the number of registers crossed while traversing different input-to-output paths depends on the chosen path.

As third and last category of elementary microarchitectures, feedback pipelines contain loops of registers as in Fig. 2.4c, as exemplified by IIR filters. In feedback pipelines, the output at the generic $n$th cycle depends on the previous inputs like the other two types of pipeline, as well as the previous outputs due to the presence of the loop(s) that feed the output back into previous pipestages. Feedback pipelines offer very limited re-pipelining opportunities because inserting or removing registers generally alters the relative timing of signals at the point of reconvergence, and

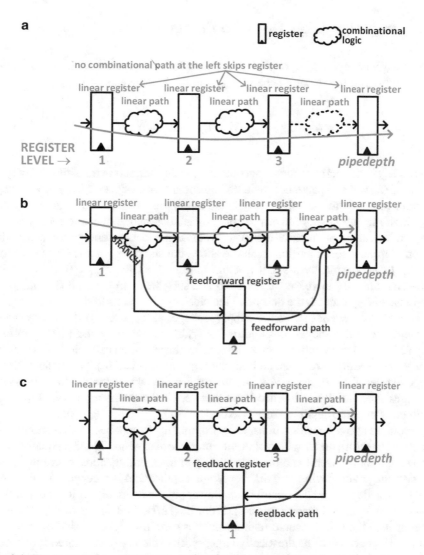

**Fig. 2.4** Elementary microarchitectures: (**a**) linear, (**b**) feedforward, (**c**) feedback pipelines

hence changes the pipeline functionality. As very few exceptions, functionality is instead preserved when inserting registers in loops through the *N*-slowing technique (see Sect. 2.6) or in unusually symmetric structures [5, 6]. Accordingly, the throughput and the energy benefits of re-pipelining can be actually extracted only if the critical path lies outside feedback loops, as fundamental property of any pipelined design.

Further details on elementary microarchitectures and the overhead of re-pipelining will be discussed in Chap. 3.

## 2.3   Impact of Logic Depth on Energy

The energy per cycle $E_{cycle}$ of digital synchronous designs consists of the two main contributions of the dynamic energy $E_{dyn}$ and the leakage energy $E_{lkg}$ [7, 8]

$$E_{cycle} = E_{dyn} + E_{lkg} = \alpha_{sw} \cdot C_{tot} \cdot V_{DD}^2 + V_{DD} \cdot I_{off,tot} \cdot T_{CK} = \alpha_{sw} \cdot C_{tot} \cdot V_{DD}^2 \left[ 1 + \frac{E_{lkg}}{E_{dyn}} \right] \quad (2.7)$$

where $\alpha_{sw}$ is the activity factor accounting for discontinuous transition activity of signals, $C_{tot}$ is the capacitance switched during the clock cycle, $I_{off,tot}$ is the leakage current drawn from the supply, $T_{CK}$ is the minimum clock period in Eq. (2.1), and $V_{DD}$ is the supply voltage. In any supply voltage region, the dynamic energy is well known to scale quadratically with $V_{DD}$, whereas the leakage energy has a very different dependence. In particular, in the above-threshold region, the leakage energy in Eq. (2.7) is typically a small fraction of $E_{cycle}$ (i.e., $E_{lkg}/E_{dyn} \approx 0$ in Eq. (2.7)), whereas it rapidly increases in the near- and sub-threshold region due to the rapid increase in $T_{CK}$. Indeed, the delay and the clock cycle in the sub-threshold regime is exponentially dependent on the supply voltage, as opposed to the nearly linear dependence in the above-threshold region [7, 8]. Accordingly, the $E_{lkg}/E_{dyn}$ ratio in Eq. (2.7) varies substantially across different voltages, often exceeding the 0.05–0.5 range across voltage regions, on both the high and low end [8]. Due to such strong dependence of $E_{lkg}/E_{dyn}$ on $V_{DD}$, the energy breakdown of the digital sub-system changes substantially and hence leads to a very different energy-optimal microarchitecture pipeline depths as discussed below.

From the above considerations, energy optimality cannot be met by a single fixed microarchitecture under wide voltage scaling. More in detail, Fig. 2.5a qualitatively shows the energy trend and breakdown of two microarchitectures operating at the same voltage and having two different logic depths. One is a deeper pipeline with lower logic depth, and the other is a shallower pipeline with higher logic depth (i.e., lower *pipedepth* and fewer registers). From Eq. (2.7), the leakage energy becomes dominant in the sub-threshold regime, and it is certainly higher in shallow microarchitectures because of its drastically larger clock cycle $T_{CK}$ (easily three-four orders of magnitude higher than at nominal voltage, or even more). Hence, deep pipelines are more energy-efficient at low voltages, when compared at iso-voltage (e.g., when $V_{DD}$ cannot be independently set for the considered block, for example, when the voltage in the relevant power domain is set by other blocks in the same domain). Also, the voltage at which the minimum energy point (MEP) is achieved in deep pipelines is pushed to the left, compared to shallow pipelines [8, 9]. In other words, deep pipelines extend the voltage range in which $V_{DD}$ scaling enables true energy reduction. Instead, shallow pipelines are more energy-efficient than deep ones at larger voltages, when compared at iso-voltage. Indeed, the larger leakage energy in shallow pipelines at above-threshold voltage has an insignificant impact on the overall energy, being dominated by $E_{dyn}$. At the same time, shallow pipelines have

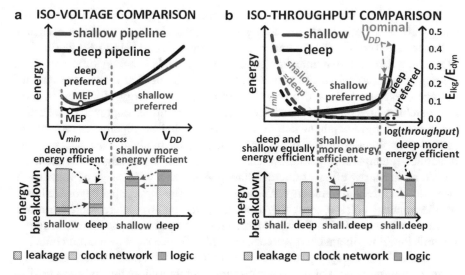

**Fig. 2.5** Qualitative trend of energy and breakdown of deep (i.e., low logic depth) and shallow (i.e., high logic depth) microarchitecture: (**a**) comparison vs. $V_{DD}$ (iso-voltage), (**b**) comparison vs. throughput (iso-throughput). The minimum energy point is labeled as MEP. In (**b**), $V_{DD}$ scaling is implicitly applied to scale the throughput. The crosspoint voltage at which the energy curves cross each other is generally different in (**a**) and (**b**)

fewer registers and hence have lower clocking energy as summarized in the qualitative breakdown in Fig. 2.5a.

Figure 2.5b shows the qualitative comparison of the energy and the breakdown of shallow and deep microarchitectures at same throughput, i.e. by individually adjusting $V_{DD}$ to equate the throughput. From a design viewpoint, this case is relevant to system architectures where the voltage of the digital block under design can be controlled independently of others. At above-threshold voltages, the throughput excess offered by deep pipelines allows for more aggressive voltage scaling, which translates into lower energy than shallow pipelines due to the dominance of $E_{dyn}$. At lower throughput targets and hence lower $V_{DD}$ in the near-threshold region, the throughput becomes much more sensitive to $V_{DD}$. Hence, the throughput disadvantage of the shallow pipeline can be fully compensated through a very limited voltage increase. In this case, the operating voltage of deep and shallow pipelines to achieve the same throughput is almost the same, and their energy essentially differs only for the lower number of registers in shallow pipelines. Since in the near-threshold region, the dynamic energy is still a major fraction of the overall budget, the lower clocking energy of shallow pipelines makes them more energy efficient than the deep ones, as shown in Fig. 2.5b. At extremely low throughput targets, the resulting voltage of deep and shallow pipelines that makes their throughput equal is again nearly the same. Since in the sub-threshold region, the energy is dominated by leakage, which is essentially the same for both, shallow and deep configurations are equally energy efficient.

From Fig. 2.5a, the adoption of a fixed microarchitecture cannot achieve the best energy efficiency at all voltages, as higher (lower) logic depth would be needed at voltages in the order of the nominal $V_{DD}$ (near-threshold and below). Similarly, from Fig. 2.5b a fixed microarchitecture cannot achieve the best energy efficiency at all throughputs, as lower (higher) logic depth would be needed around the nominal voltage (near-threshold and below). Accordingly, the energy efficiency of fixed microarchitectures under wide voltage scaling can be further improved by adjusting the pipestage logic depth, switching from deep to shallow pipeline as required by the adopted voltage or the throughput target. This motivates the necessity for dynamically adaptable pipelines, where the logic depth can be adjusted as needed to have a more energy efficient microarchitecture across a wide range of voltages. In addition, such reconfiguration permits to improve the throughput beyond allowed by the nominal voltage, switching from a slower shallow microarchitecture to a faster deeper one as in Fig. 2.5b. Also, reconfiguration extends the minimum energy towards lower minimum energy points, switching from a shallow microarchitecture with higher MEP energy to a deeper one with lower MEP energy as in Fig. 2.5a. As further expected benefit of microarchitecture reconfiguration, the ability to flexibly select the deep pipeline moderates the rapid energy increase below the MEP voltage of shallow pipelines (see Fig. 2.5a), thus making the energy around the MEP flatter, and hence relaxing the accuracy requirement of the supply voltage when minimum energy is targeted.

## 2.4  Dynamically Adaptable Pipelines

In this section, dynamically adaptable pipelines are introduced to enable energy reductions via joint microarchitectural and voltage co-optimization, for voltages widely ranging from nominal voltage down to deep sub-threshold. This section is structured as follows. In Sect. 2.4.1, the basics on wide voltage scaling are briefly covered. Then, the concept of dynamically adaptable pipelines is introduced in Sect. 2.4.2, and their incorporation into conventional schemes for dynamic voltage scaling is discussed in Sect. 2.4.3.

### 2.4.1  Wide Dynamic Voltage Frequency Scaling

Wide dynamic voltage frequency scaling (DVFS) has been extensively adopted to reduce the energy consumption, whenever the performance target lies in a wide range below the maximum achievable [10–17]. The energy savings due to DVFS at above- and near-threshold supply voltages are nearly quadratic due to the dominance of the dynamic energy in Eq. (2.7). Within the same voltage range, the power savings are approximately cubic [7, 8]. At lower voltages, further energy reductions are achieved although the returns diminish when $V_{DD}$ approaches the minimum energy point (MEP), due to the increased contribution of the leakage energy in

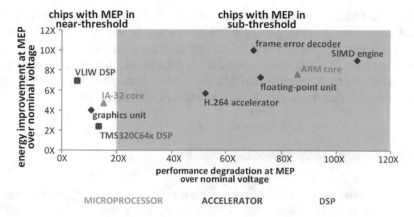

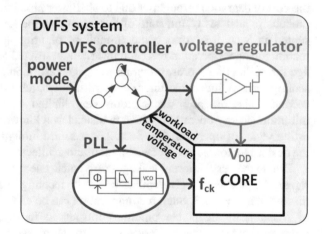

**Fig. 2.6** Energy-performance improvement over nominal voltage under wide voltage scaling from [8]. The *y*-axis shows the energy gain when $V_{DD}$ is reduced from nominal voltage down to MEP, whereas the *x*-axis shows the corresponding throughput penalty due to the same voltage down-scaling

**Fig. 2.7** Simplified dynamic voltage frequency scaling scheme

Eq. (2.7) [18, 19]. As exemplified in Fig. 2.6 with data taken from [8], wide voltage scaling permits to achieve 2–6× energy reduction when the MEP is at near-threshold voltages and 10× when the MEP is in sub-threshold, compared to nominal voltage. This is achieved at the cost of a throughput degradation by up to several tens of times compared to nominal voltage.

DVFS is implemented as a look up table that associates each supply voltage to the corresponding maximum clock frequency that meets the timing constraints discussed in Sect. 2.1.1, as shown in Fig. 2.7. The table entry indicating the supply voltage is fed to the voltage regulator that powers the voltage-scaled domain. The entry associated with the clock frequency is fed to the on-chip clock generator (e.g., phase-locked loop) that drives the clock distribution network of the block utilizing DVFS. The voltage–frequency pair is chosen according to the power mode (usually set at the software or firmware level), which in turns is defined by the targeted throughput.

Communication between the core and the DVFS controller in Fig. 2.7 is bidirectional. Indeed, the processor sends information about its environmental parameters such as the temperature and voltage, to enable voltage–frequency fine-tuning and timing margin elimination due to process, voltage, and temperature variations. Environmental sensors are typically implemented by multiple distributed ring oscillators, whose frequency is used to extrapolate such variations [18].

## 2.4.2  Dynamically Adaptable Pipeline

Dynamic pipeline stage unification has been previously investigated as form of microarchitectural reconfiguration at the pipeline stage level to save energy by reducing the pipeline depth at nominal voltage [20–25], or over a narrow range of above-threshold voltages [26]. In these simple cases, a reduced pipeline depth invariably leads to fewer registers (i.e., lower clocking energy) and less frequent hazards (i.e., lower energy per instruction) in microprocessor systems, although at the cost of degraded throughput due to shallower pipelining [20–26]. A more comprehensive analysis of the state of the art in pipeline-level reconfiguration is presented in Sect. 3.1. Among the limitations of prior art on pipeline stage-level reconfiguration, operation has been invariably restricted to a fixed or a narrow voltage range. In other words, previous work ignores the implications of wide voltage scaling, and its inevitable interaction with microarchitectural reconfiguration. As second limitation, prior art explores very limited energy gains while ignoring minimum-energy operation, which is instead well known to be a particularly interesting region of operation (see Sect. 2.3). As third limitation, no prior art addresses the design methodology aspects of joint microarchitectural and wide voltage scaling.

Due to the very different leakage-to-dynamic energy ratio across a wide voltage range, the pipestage logic depth needs to be reconfigured at different voltages, as discussed in Sect. 2.3. Such reconfiguration can be enabled by the general scheme of dynamically adaptable pipelines introduced in Fig. 2.8. From this figure, microarchitecture reconfiguration down to the pipestage level is enabled by the introduction of bypassable registers, each of which can operate in either bypass or normal mode, depending on the value of the *bypass* signal. In bypass mode, *bypass* is set to 1 so that the multiplexers embedded in these registers can bypass the corresponding flip-flops, and the shared clock gater disables their clock to suppress their power. These multiplexers effectively merge the previous and the subsequent pipestage into a single one, thus increasing the logic depth as desired. In normal mode, *bypass* is set to 0 and the multiplexer selects the flip-flop output, while the clock gater enables the clock. This leads to conventional register operation, and the selection of the smallest available logic depth.

The dynamically adaptable pipeline scheme in Fig. 2.8 does not modify the clock network, and hence maintains the same clock load in all modes, as shown in the adopted clock tree architecture in Fig. 2.9. This is because the register bypass is entirely performed in the datapath and the load is not altered in the clock path,

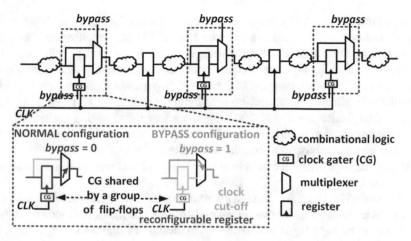

**Fig. 2.8** Dynamic adaptable pipelines are enabled by the introduction of bypassable registers. The latter ones can operate either in bypass mode (*bypass*=1) to merge the previous and the subsequent pipestage into one or in normal mode (*bypass*=0) to conventionally keep the two pipestages separate. In this example, bypassable registers are placed in odd-numbered pipestages as a highly representative example and can be alternatively placed according to the patterns discussed in Fig. 2.10a–d

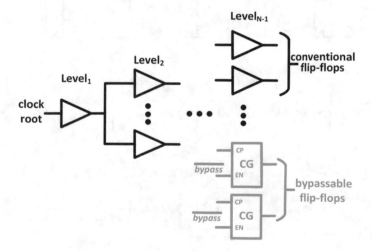

**Fig. 2.9** Clock tree architecture in dynamically adaptable microarchitecture

whereas the only change across configurations is the selective activation of clock gaters. Accordingly, the clock skew is independent of the selected microarchitecture configuration. This eliminates the risk of introducing additional setup/hold time violations when switching to a different configuration, providing a guarantee on the timing closure (i.e., microarchitecture reconfiguration does not induce any timing error).

As a generalization of conventional static pipelined designs, the delays across reconfigurable pipeline stages need to be balanced in both shallow and deep configuration, as necessary to preserve their throughput and energy efficiency. This generalized balanced pipestage requirement permits to rapidly determine the few feasible combinations of bypassable and non-bypassable registers that should be considered in the microarchitecture reconfiguration. This small pool of combinations is then further narrowed down by discarding those having an unacceptable area or energy cost of bypassing the selected registers (e.g., if the bypassable registers in a given combination account for a substantial fraction of the overall gate count). In the following, an example of how the feasible combinations are identified is presented.

Let us consider the six-pipestage microarchitecture in Fig. 2.10a. As usual, deep configuration is obtained by making all registers operate in normal mode and having the delays of all pipestages well balanced by design. In shallow configuration, any of the five registers in Fig. 2.10a can be made bypassable, and in principle, a large number of combinations of bypassable and non-bypassable registers exist. However, only a few of such combinations are able to be balanced in the shallow

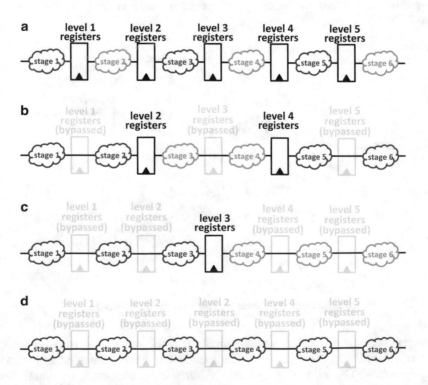

**Fig. 2.10** Feasible combinations of bypassable and non-bypassable registers keep pipestages balanced in all configurations. (**a**) In deep configuration, all $n = 6$ pipestages are balanced by design. Balanced shallow configurations merge $i$ adjacent pipestages of the deep configuration, with $i$ being a divisor of $n$: (**b**) $i = 2$, (**c**) $i = 3$, (**d**) $i = 6$

configuration(s), once the original pipestages in the deep configuration are well balanced as above. From the example in Fig. 2.10a–d, there are only three possible options that keep stages balanced when registers are bypassed:

1. Registers at level 1, 3, 5 are made bypassable, so that pairs of adjacent pipestages are merged in the shallow configuration (i.e., #1–2, #3–4, and #5–6)
2. Registers at level 1, 2, 4, 5 are made bypassable, so that triplets of adjacent pipestages are merged in the shallow configuration (i.e., #1–3, #4–6)
3. All registers are made bypassable, so that all pipestages are merged into a single one in the shallow configuration.

From the above example, it is clear that the feasible combinations with balanced pipestages in the shallow configuration are achieved only by grouping and merging every number $i$ of consecutive pipestages of the deep configuration, with $i$ being a divisor of the number $n$ of available pipestages (i.e., $i = 2$, 3, and 6 as divisors of $n = 6$, from Fig. 2.10a–d). Then, the intermediate registers in each merging group are made bypassable in each combination. The selection of the most appropriate combination is discussed in the following.

Among the feasible bypassable register combinations, some might be then discarded because of an unacceptably large area/energy cost of register bypassing. To make the above example more quantitative, post-PNR results in 40 nm are provided for a six-pipestage computational element used inside an FFT accelerator (i.e., multipliers and adders), which is discussed in detail in Chap. 4. In this example, the cost of register bypassing in the combinations in Fig. 2.10b–d is characterized in the following:

1. *Merge every i = 2 consecutive registers.* In this case (Fig. 2.10b), the logic depth in shallow configuration is approximately twice the logic depth of the deep configuration (Fig. 2.10a). Compared to the fixed deep configuration (Fig. 2.10a), pipestages in Fig. 2.10b occupy slightly larger area as bypassable registers inserted in level 1, 3, and 5 are more complex than conventional registers. In detail, 9200 flip-flops are made bypassable out of a total of 29,100 flip-flops, resulting into 3.4% larger area and 3.4% larger leakage power, compared to the fixed deep configuration. Also, the clock cycle increases by 2.05×, and the energy leakage increases by slightly more (2.11×) due to the slight increase in the leakage power.
2. *Merge every i = 3 consecutive registers.* In this case (Fig. 2.10c), the logic depth in shallow configuration is approximately 3× larger than the deep configuration (Fig. 2.10a). The replacement of regular registers by the bypassable type leads to a 7.4% area and 7.6% leakage power increase, compared to the fixed deep configuration. Also, the energy leakage increases by 3.32×, which is close to the 3.1× clock cycle increase due to the rather limited leakage power increase.
3. *Merge all registers (i = 6).* The logic depth in Fig. 2.10d is expectedly approximately six times larger than the fixed deep configuration. Compared to the latter, this combination occupies 10.4% larger area due to the presence of five levels of bypassable registers, as opposed to three (four) levels in Fig. 2.10b (Fig. 2.10c).

**Table 2.1** Area and leakage energy/power penalty of register bypassing (six-pipestage computational element for FFT in Chap. 4)

| | Fixed deep | Reconfigurable shallow | | |
| --- | --- | --- | --- | --- |
| | | $i = 2$ (Fig. 2.10b) | $i = 3$ (Fig. 2.10c) | $i = 6$ (Fig. 2.10d) |
| # Bypassable flip-flops/total | – | 9200/29,100 | 23,200/29,100× | 29,100/29,100× |
| Clock cycle | 1× | 2.05× | 3.1× | 6.25× |
| Area | 1× | 1.034× | 1.074× | 1.104× |
| Leakage power | 1× | 1.034× | 1.076× | 1.101× |
| Leakage energy | 1× | 2.11× | 3.32× | 6.87× |

For the same reason, the leakage power of the configuration shown in Fig. 2.10d increases by 10.1% over Fig. 2.10a. The 6.87× energy leakage increase is again close to the 6.25× clock cycle increase, due to the limited leakage power increase.

The above results are summarized in Table 2.1 and indicate that the area cost of register bypassing is reasonable in all combinations in the considered example. Similar considerations hold in terms of leakage power penalty, which is fairly small as the leakage energy increase is still basically determined by the clock cycle increase in Eq. (2.7), rather than the increase in the leakage power. Accordingly, all three combinations can be retained as feasible candidate designs, and their choice is made based on further considerations on the tradeoff between flexibility and the energy benefit, as discussed below.

It is worth noting that the flip-flop count in the considered design is an uncommonly large fraction of the gate count (24% from Table 4.1). Hence, the above considerations are expected to hold in more common designs with a lower gate count percentage spent on flip-flops, keeping the area and the energy reconfiguration cost moderate. Similarly, the leakage energy in common and less register-intensive designs is basically proportional to the clock cycle and hence to the number of consecutively bypassed pipestages $i$ in the shallow configuration (since all merged pipestages have nearly the same delay, due to the initial delay balancing in the deep configuration).

In a dynamically adaptable microarchitecture, maximum flexibility in terms of achievable logic depth would be obtained by making all registers bypassable. However, this would be impractical as the cost of bypassable registers would be the highest, adding both a delay and an energy overhead due to their embedded multiplexer. Also, this would make pipeline stage balancing more difficult for commercial EDA tools performing retiming, thus increasing the delay overhead. In line with the above example, experiments with several microarchitectures (see Chap. 4) showed that such overhead is reasonably small and the microarchitecture is nearly energy-optimal when bypassable registers are placed at every other pipestage (i.e., $i = 2$ and the logic depth can be hence increased by a factor of approximately 2×, compared to the fixed deep configuration). In other words, a reasonable balance between microarchitectural flexibility and energy benefits is to insert bypassable registers only in odd-numbered register levels as in Fig. 2.8, while leaving even-numbered registers as conventional registers.

### 2.4.3 Run-Time Pipeline Adaptation via Augmented DVFS Look-Up Table

The above-discussed microarchitectural adaptation to the voltage and/or the throughput target at run-time can be straightforwardly integrated with conventional DVFS schemes (see Fig. 2.7). Indeed, the analysis in Sect. 2.3 showed that each of the shallow and deep configurations in Fig. 2.5a, b is preferred in a voltage range going from the crosspoint voltage $V_{cross}$ to the nominal voltage (or vice versa), regardless of the specific design and targets. In other words, for a given voltage the microarchitectural configuration is unambiguously defined, along with its maximum clock frequency.

According to the above considerations, the scheme in Fig. 2.11 can be adopted to jointly scale the voltage, the clock frequency, and the most energy-optimal microarchitecture configuration. In other words, the only necessary modification in conventional DVFS schemes to incorporate the above microarchitecture reconfiguration is the enhancement of the voltage–frequency look-up table with the definition of the microarchitecture configuration for each voltage. In the nearly optimal scheme in Fig. 2.8, this is simply achieved by adding the appropriate value of *bypass* to an additional column in the DVFS look-up table. In this way, operation at a given voltage (see Fig. 2.5a) sets both the clock frequency and the logic depth of the microarchitecture (i.e., deep if *bypass* = 0, shallow if *bypass* = 1), as desired. Alternatively, operation at a given throughput (see Fig. 2.5b) sets the voltage and the microarchitecture.

Regarding the microarchitecture configuration choice at run-time, the above scheme assumes that the look-up table is pre-compiled to maximize the energy efficiency across a representative set of workload benchmarks (as assessed at design or testing time). Alternatively, energy efficiency can be further improved by adapting the crosspoint voltage $V_{cross}$ to the specific workload being executed, when the volt-

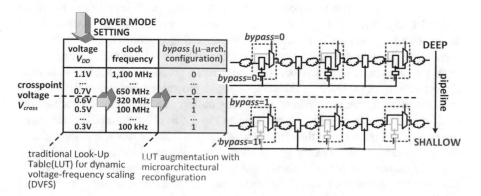

**Fig. 2.11** Integration of microarchitecture configuration in conventional DVFS schemes and run-time co-optimization under a given $V_{DD}$ (i.e., as derived from considerations in Fig. 2.5a). The choice of deep/shallow configuration is swapped when $V_{DD}$ is optimized to achieve a given throughput (as derived from considerations in Fig. 2.5b)

age regulator powering the module under design has an embedded power sensor. For simplicity, the former approach has been adopted throughout the examples in this book.

## 2.5  Microprocessor Microarchitectures: Opportunities and Challenges Under Reconfiguration

### 2.5.1  Wide DVFS in Microprocessors and Considerations at the Application Level

Microprocessors with wide energy–performance scalability are of great interest in several applications. As a representative example, self-powered platforms (e.g., via a battery and/or an energy harvester [8]) are tightly constrained in terms of power, and hence require nearly minimum energy operation in the common case, under the assigned throughput requirement (usually very relaxed to keep power low). Low-energy operation is particularly critical in the case of always-on systems that continuously monitor data coming from sensors, in search for events of interest [27, 28] (indeed, power is given by the product of energy and throughput, and the latter is set by the application or operating mode). At the same time, nowadays emphasis is being given to edge computing to further limit the power penalty of wireless communications, and the inevitable data deluge created by the transmission of raw data [8]. Accordingly, the data produced by sensors under event occurrence is processed at the edge rather than in the cloud, thus requiring both event detection and rapid response through significant on-chip data analytics for nearly real-time data sensemaking [27, 28]. Hence, edge computing need to deliver substantial extra peak performance to promptly respond to events upon their occurrence [10, 29]. This has motivated a general push towards increasingly wide energy-performance scaling.

From an application viewpoint, occasional throughput enhancement might be needed to cope with the higher processing demands from a larger number of sensor channels, e.g., when more accurate data sensemaking is required. For similar reasons, higher throughput could be occasionally needed to sustain higher sample rates from the same number of sensors. In either case, a processor with a fixed microarchitecture cannot increase its throughput above the value obtained at the nominal voltage. Instead, microarchitecture reconfiguration can exceed such throughput limit at the cost of higher energy, as shown in Fig. 2.5b and as exemplified in operating point 1 in Fig. 2.12. At the same time, microarchitecture reconfiguration allows to reduce the minimum energy, as shown in Fig. 2.5a and as indicated by point 2 in Fig. 2.12. In other words, microprocessor microarchitecture reconfiguration extends the power-performance tradeoff beyond mere voltage scaling, as was observed for application-specific accelerators in Sect. 2.3.

Although microarchitecture reconfiguration is potentially highly beneficial, its enablement through re-pipelining in processor microarchitectures is difficult and entails several drawbacks that need to be overcome, as discussed in the next two subsections.

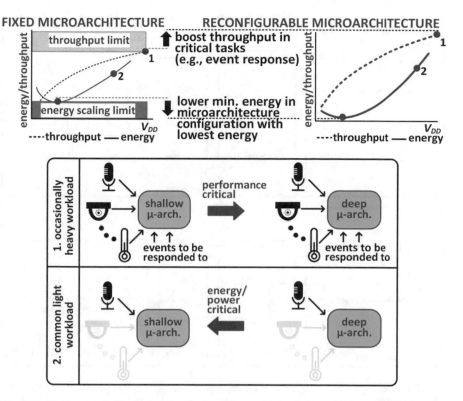

**Fig. 2.12** Processor microarchitecture reconfiguration and joint co-optimization with supply voltage extends the power-performance tradeoff (top) and permits to improve the energy efficiency both at the higher and the lower end of the throughput range (bottom of the figure, based on the energy-optimal configuration choice from Fig. 2.5a, b). This creates additional opportunities to save power when the workload dynamically changes

## 2.5.2  Control Flow and Hazards in Microprocessor Microarchitectures with Different Pipedepths

Figure 2.13 shows a typical six-pipestage RISC processor with Instruction Fetch, Instruction Decode, and Execute stages. In the same figure, the five-pipestage re-pipelined version is shown with pipestages being Instruction Fetch, Decode, Execute, Memory, and Write Back. Compared to the three-pipestage version, the five-pipestage processor control flow sometimes needs to add *pipeline bubbles* (no-operation instructions) in the pipeline, to avoid hazards due to data inter-dependency.

As an example, Fig. 2.14 shows the assembly code where the value of register #3 is first added with the constant value 6 and stored in register #0. In the second instruction, the updated value of register #0 is read, added to register #4, and stored in register #3. The third instruction adds the constant 2 to the updated register #3

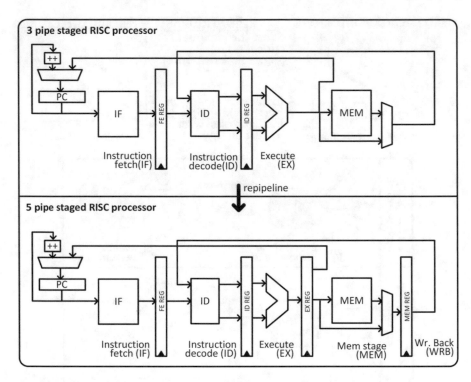

**Fig. 2.13** Three-pipestage RISC pipelined microprocessor (top) and re-pipelined five-pipestage version (bottom)

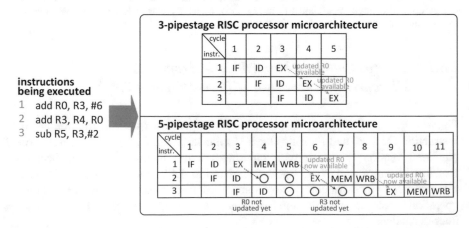

**Fig. 2.14** Instruction execution vs. cycle and pipeline bubble insertion when data inter-dependency prevents an instruction from being immediately executed, when it needs to wait for the completion of the previous one

value, and stores it in register #5. In the three-pipestage microarchitecture, the Execute stage computes the addition (instruction #1) and stores it in the register #0 within the same cycle. When the Execute stage is storing data in the register #0 for instruction #1, the Decode stage is decoding instruction #2 from Fig. 2.14, and the updated value of register #0 from instruction #1 at cycle #3 can be used by instruction #2 right after at cycle #4. On the other hand, in a five-pipestage microarchitecture, the write back onto registers occurs only after five cycles from the instruction fetch; hence, their updated value is available to the next instruction only at this cycle. This creates a data inter-dependency between instructions accessing registers just updated by the previous instruction. For example, register #0 is written by instruction #1 at cycle #5 in Fig. 2.14; hence, instruction #2 that requires the updated value of register #0 has to wait until cycle #6, instead of being executed at cycle #4 from Fig. 2.14. Accordingly, the control flow inserts two stall cycles, which arc represented as blue pipeline bubbles in Fig. 2.14. Similarly, instruction #3 should read register #3 at cycle 5, but actually this needs to wait for the updated register #3 from instruction #2, which is available at its write back stage only right after cycle #8 (i.e., at cycle #9). Hence other two (red) bubbles are inserted in instruction #3 in Fig. 2.14.

From the simple example in Fig. 2.14, the control flow overhead and the effect of data inter-dependencies can heavily increase in deeply pipelined structures, and the penalty of re-pipelining and hence microarchitecture reconfiguration can be heavy.

### 2.5.3   Limitations of Re-pipelining in Existing Microprocessor Architectures

The reconfiguration approach discussed in the previous sections introduces re-pipelining into a baseline deep microarchitecture design. However, re-pipelining is hard to be automated in microprocessors because of several reasons. First, a change in the processor pipeline requires a redesign of the software compiler, and hence substantial effort for software re-development and re-debugging, as well as legacy issues. The adaptation of the compiler to multiple microarchitectures (as required by reconfiguration) makes its design even harder.

Second, the control flow of the processor would need to be redesigned to accommodate the modified pipeline structure, and comprehensively deal with all possible additional microarchitectural hazards (see, e.g., example in Sect. 2.5.2). Again, this requires a redesign of the processor architecture and extensive verification of its implementation.

Third, even if the processor architecture with modified and reconfigurable control flow were made available, a deeper pipeline would induce more stalls than shallower pipeline (especially if the baseline has an inherently deep pipeline). This would in turn degrade the performance and the potential energy efficiency benefits coming from reconfiguration.

Fourth, SRAM caches of processors continuously interact with the processor and are usually in its critical path (e.g., instruction fetch stage). Therefore, even though the processor is made faster through deep pipelining, the throughput would not be improved if the associated memories (both program and data memory) were not made fast enough. The presence of the SRAM memory in the data path of processor also restricts the retiming process. Hence, SRAMs need to be made consistently faster when the processor is reconfigured for higher throughput. This requires either substantial SRAM overdesign (i.e., at extra area and energy cost) to keep up with the maximum processor throughput or reconfiguration in the SRAM itself.

Finally, the inevitable presence of loops in microprocessors (see, e.g., the loops in Fig. 2.13) prevents re-pipelining and restricts the movement of flip-flops during retiming and pipestage re-balancing. Due to all these reasons, microarchitecture reconfiguration in general-purpose processors cannot be performed through simple re-pipelining, and an alternative approach needs to be followed, as discussed in the next section.

## 2.6   Enabling Microarchitectural Reconfiguration in Microprocessors

This section introduces parallelism and time interleaving as microarchitecture reconfiguration alternatives to re-pipelining.

Parallelism inherently enables microarchitecture reconfiguration by activating different number of cores based on the targeted power-performance tradeoff, as schematized in Fig. 2.15. However, scaling up the throughput of a single processing unit has a very substantial area overhead, which is proportional to the degree of parallelism. This might be justified only in high-performance applications, whereas a more favorable throughput-area tradeoff can be enabled by time interleaving, as discussed in the following.

Time interleaving permits to reuse the same processing element to compute $N$ independent time-interleaved input data streams and deliver their corresponding outputs in a time-interleaved manner [5, 6, 30]. In general, a time-interleaved signal is a single stream of data merging $N$ streams in such a way that the $i$-th piece of data

**Fig. 2.15** General scheme of parallelism with $N$ replicas of the processing element (e.g., microprocessor core), each driven by a separate input stream

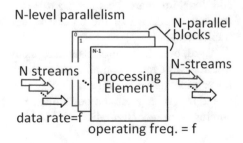

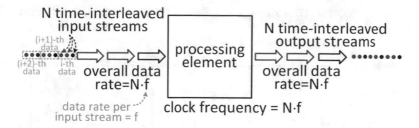

**Fig. 2.16** Time-interleaved system with $N$ input streams and output streams, representing channels coming from different information sources. The generic $i$-th piece of time-interleaved data contains the corresponding value for streams 1, 2...$N$

of the time-interleaved stream consists of the ordered set of the $i$-th piece of data of streams 1, 2...$N$, as shown in Fig. 2.16.

Then, the same sequence is repeated for the $(i + 1)$-th piece of data, and so on, as in Fig. 2.16. From this figure, a time-interleaved design needs to operate at a clock frequency that is $N$ times larger than the data rate of each input stream. In case the $N$ input streams represent physically independent channels (e.g., data from independent sensors), their composition into the time-interleaved stream requires an additional parallel-to-serial interface at the input. Similarly, a serial-to-parallel interface is needed to generate the independent output streams from the time-interleaved one [5, 6].

Interestingly, a conventional microarchitecture with a single input stream can be turned into a time-interleaved one merging $N$ streams via a very simple manipulation of its registers. More in detail, it is sufficient to start from the conventional single-input microarchitecture, and replace each register by $N$ cascaded registers as in Fig. 2.17 [5, 6]. At a given point of time, each group of $N$ cascaded registers holds the values related to the $N$ streams for a given piece of data (input, output, or any intermediate signal). Hence, the output related to a given stream is updated every $N$ cycles, i.e., its computation is completed at a frequency that is $N$ times slower than the clock frequency, as expectable from any time-interleaved system [5, 6].

As interesting property, time interleaving permits to process $N$ streams rather than one by simply replicating the registers, while keeping the complexity of the combinational logic exactly the same. In other words, the area and energy cost of time interleaving is relatively low as it only replicates the registers, as opposed to registers and combinational logic in parallelism.

The above register manipulation in Fig. 2.17 enables the independent processing of separate input streams, but actually degrades the throughput per channel by a factor of $N$. This is because the number of cycles to complete a computation is multiplied by $N$ compared to the single-stream system, whereas the clock frequency remains exactly the same. In turn, this is due to the fact that the critical path does not change, and the fast register-to-register paths between cascaded registers do not affect the worst-case clock frequency across pipestages anyway.

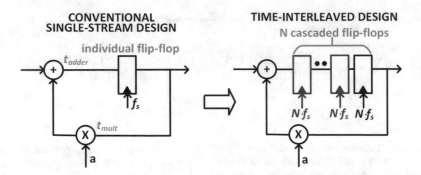

**Fig. 2.17** The gate-level design of a conventional single-stream system is translated into a time-interleaved one with $N$ streams by replacing each register (left) with $N$ cascaded registers (right)

Retiming of the time-interleaved version of the microarchitecture permits to eliminate the above degradation in the throughput per channel, compared to the original single-stream microarchitecture. Indeed, retiming the cascaded registers in Fig. 2.17 permits to ideally distribute the same combinational logic depth across $N$ adjacent pipestages, reducing the logic depth per pipestage and hence increasing the clock frequency by a factor of $N$ [5]. Hence, the throughput for each stream after retiming is slower than the original one by a factor of $N$ due to time interleaving, but it is accelerated by nearly the same factor through the clock frequency increase compared to the original design. Thus, the throughput per channel is the same as the original single-stream design, and the overall throughput across the $N$ streams is hence nearly $N$ times larger than the original design. This substantial throughput improvement is achieved at a moderate register area and energy cost, due to the replication of registers only.

In summary, time interleaving permits to increase the throughput and reduce the pipestage logic depth to sustain $N$ independent streams, while maintaining the same original throughput for each stream. This technique is very general, can be applied to any microarchitecture (even including feedback loops [6]), and is also easy to apply as it simply requires the replacement of each register by $N$ cascaded registers. Given its limited area and energy overhead associated with registers only, while effectively turning a single core into $N$ independent cores, time interleaving is a very interesting technique to modify the pipeline structure of microprocessors. Instead of applying time interleaving in a static microarchitecture, it can also support reconfiguration as discussed in the following section.

## 2.7 Dynamically Adaptable Time-Interleaved Microprocessors

Time interleaving in a microprocessor allows to improve the overall throughput by scaling up the number of independent instruction streams (named "threads"), when necessary. This suggests that selective activation of time interleaving provide another mean to extend the power-performance tradeoff, introduce joint microarchitecture and voltage co-optimization at run-time, similar to the reconfigurable re-pipelining discussed in Sect. 2.4.2. Indeed, in essence the activation of time interleaving can dynamically reduce the logic depth, whereas its deactivation via register bypassing pushes the logic depth to higher values.

According to the above considerations, a very similar mechanism as in Sect. 2.4.2 can be introduced for time interleaving, introducing $N - 1$ additional cascaded registers to each existing register as in Fig. 2.17, and making them selectively bypassable. This enables the adjustment of the number of threads being executed and hence thread-level reconfiguration. This is different from the pipeline-level reconfiguration in Sect. 2.4 that directly adjusted the pipeline structure seen by the data flow, as no change is effectively seen by each thread in terms of microarchitecture, once each separate output stream is considered (see below). When no register is bypassed, the maximum thread count $N$ and the highest overall throughput are achieved (P+ mode), making the single core behave like $N$ independent "virtual" cores. When registers are properly bypassed (see feasible combinations in Sect. 2.4.2), the number of threads is scaled down. The lowest throughput is given equal to the original microarchitecture (P mode), and is reached when all inserted registers are bypassed.

In general, a larger number of threads $N$ is supported by inserting a larger number of registers $N$. Similar to re-pipelining, the number of inserted registers is upper bounded by considerations on the logic depth, due to the diminishing clock cycle reduction returns obtained at excessively low logic depths in any pipelined microarchitecture (see Sect. 2.1.1). Typically, the maximum number of threads is set by the ratio of the logic depth of the original microarchitecture and the minimum logic depth below which diminishing returns are observed (e.g., in the range of 10–20 FO4 from Sect. 2.1.1). For example, an ARM Cortex-M0 with a logic depth of approximately 110 FO4 allows a number of threads in the order of several units (e.g., up to 6–8). On the very opposite end of the microarchitecture spectrum, high-performance processors with logic depths of 20 FO4 or less offer no opportunity to further introduce time interleaving (they already embed some form of multi-threading anyway). Typical intermediate microarchitectures with a few tens of FO4 per stage allow a few threads to be executed simultaneously.

Thread-level reconfiguration is exemplified in the processor microarchitecture example in Fig. 2.18a–d. The microarchitecture in Fig. 2.18b was derived from the three-pipestage RISC processor microarchitecture in Fig. 2.18a by doubling every register as in Fig. 2.17, making it dual-thread (i.e., time interleaved with two input streams). As discussed in Sect. 2.6, retiming is then applied to generate the microarchitecture in Fig. 2.18c, which re-balances the stage delays and regains nearly the same throughput per channel as the original microarchitecture in Fig. 2.18a. As a result, every pipestage in the original microarchitecture in Fig. 2.18a is split into two cascaded pipestages with halved logic depth. For example, IF is split into the two pipestages $IF_1$ and $IF_2$ in Fig. 2.18c, and similar considerations apply to the ID and EX pipestages. As discussed in the previous section, the pipestages $IF_1$, $ID_1$ and $EX_1$ execute thread #1 at odd-numbered cycles, while $IF_2$, $ID_2$, and $EX_2$ execute the thread

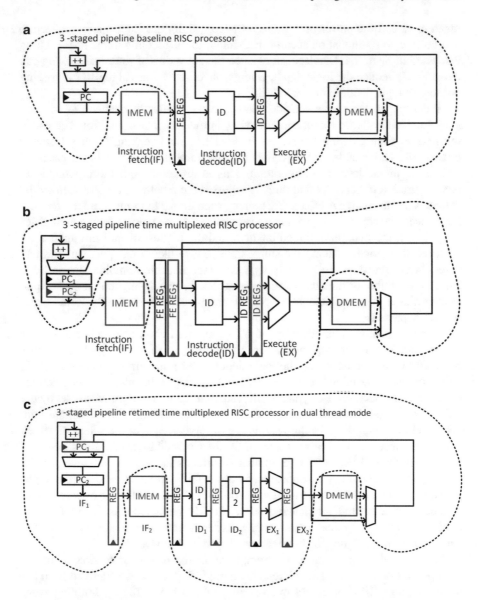

**Fig. 2.18** (**a**) Three-pipestage baseline RISC processor with fetch, decode, and execute stages, (**b**) time-interleaved version of the three-pipestage RISC processor, as obtained by replacing all registers with two cascaded registers (including register files), (**c**) retimed thread-level reconfigurable three-pipestage RISC processor microarchitecture operating in dual-thread mode P+ (all registers in **a**) were doubled, and none of them is bypassed, (**d**) same thread-level reconfigurable microarchitecture operating in single-thread mode P (even-numbered registers are bypassed for conventional operation as in (**a**))

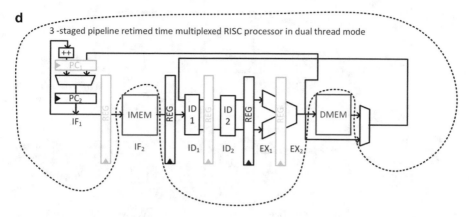

**Fig. 2.18**  (continued)

#2. The opposite occurs at even-numbered cycles, when the second half operation of each pipestage is executed to complete the full pipestage operation in two cycles, as expected from time-interleaved systems.

Once registers are properly made bypassable, dual-thread processor operation in P+ mode takes place when the dual-thread signal DT is set to 1. As shown in Fig. 2.18a, no register is bypassed and time interleaving takes place. Instead, normal (single-thread) processor configuration in P mode is set when the dual thread signal DT is assigned to 0. In this case, even-numbered register levels are bypassed to assure the same operation as the baseline microarchitectures (see registers in gray in Fig. 2.18b). In this case, the bypassable registers are clock gated to suppress their energy consumption since they are not utilized, and all registers share the same clock network. The design methodology to identify the registers to be made bypassable will be discussed in the next chapter.

Figure 2.19 shows the detailed instruction flow in the processor pipeline in Fig. 2.18a, b in both configurations. In dual-thread (P+) mode, the instructions from each thread are time interleaved and executed at nearly doubled clock frequency compared to the original microarchitecture in Fig. 2.13 (see previous section). This factor of two is slightly degraded due to the multiplexer timing overhead and imperfect cell sizing and pipestage balancing of commercially available retiming tools. Each instruction is completed in six cycles (provided there are no stalls) at nearly doubled frequency, thus maintaining the same time spent for each instruction, and the same throughput as the original microarchitecture in Fig. 2.13. In single-thread P mode, the processor runs instructions every cycle, with each instruction being completed in three cycles (provided there are no stalls).

Overall, thread-level reconfiguration brings to processors the same advantages of re-pipelining in application-specific microarchitectures, in view of the common capability of dynamically adjusting the logic depth. Hence, the analysis of the benefits of thread-level microarchitecture reconfiguration on energy and throughput is

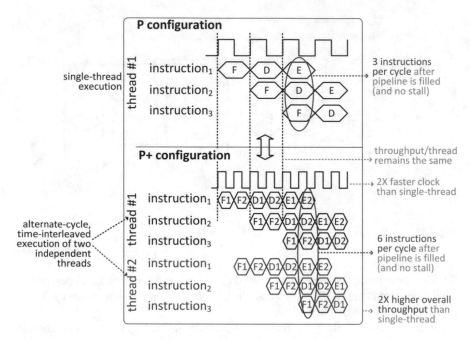

**Fig. 2.19** Instruction execution in single-thread (P) and dual-thread mode (P+)

analogous to Sects. 2.3 and 2.4. In terms of comparison at iso-voltage, the general considerations on pipelining in Fig. 2.5a also hold for thread-level reconfiguration, as shown in Fig. 2.20. In particular, the dual-thread P+ configuration is more energy-efficient at lower supply voltages (i.e., near- and sub-threshold), due to the lower leakage energy enabled by the nearly halved clock cycle. At higher supply voltages, the single-thread P configuration is instead preferred, thanks to its lower clocking and dynamic energy. Opposite considerations expectedly hold for the iso-throughput comparison, which is similar to Fig. 2.5b and is omitted in Fig. 2.20 for the sake of brevity.

In addition to the benefits brought by logic depth adjustment as in re-pipelining, thread-level reconfiguration is very well suited for general-purpose microarchitec-tures since it solves all the traditional issues of re-pipelining pointed out in Sects. 2.5.2 and 2.5.3. First, no change in the software stack is needed since each thread is processed exactly like the original single-thread microarchitecture, once the output of each thread is extracted from the time-interleaved stream (i.e., sub-sampling by a factor of $N$). Hence, the same compiler, operating system, and binary code of applications are entirely reused, solving the legacy issues of re-pipelining. Also, microarchitectural adaptation is inherently dealt with by the compiler, as each thread is simply compiled independently and mapped to different memory sub-spaces.

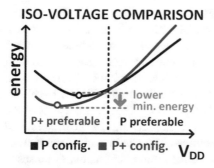

**Fig. 2.20** Qualitative trend of energy in dual-thread and single-thread processor microarchitecture and comparison vs. $V_{DD}$ (iso-voltage). This trend is expectedly similar to re-pipelining in Fig. 2.5a, due to the common fact that the logic depth is being adjusted. An analogous plot (with swapped roles for P and P+) is obtained at iso-throughput, similarly to Fig. 2.5b (omitted here, for brevity)

Second, no change in the processor control flow is needed, as time interleaving inherently prevents any interaction between instructions, and the hazards in each thread is dealt with exactly in the same way as a single thread.

Third, the pipeline structure seen by each instruction is exactly the same as the original microarchitecture; hence, no performance and energy efficiency degradation due to stalls is added to the latter. Finally, time interleaving and its reconfiguration is applicable to any synchronous microarchitecture including feedback loops, as opposed to re-pipelining. Accordingly, it is a general approach that can be applied to any processor architecture, regardless of the inevitable presence of loops.

In summary, thread-level reconfiguration brings all the benefits of logic depth adjustment and joint co-optimization with the supply voltage, while not suffering from the limitations of re-pipelining in general-purpose microarchitectures. Also, thread-level reconfiguration avoids any change in the software stack and the processor architecture, enabling full software and processor architecture reuse and minimal invasiveness. In addition, the thread-level approach is a drop-in solution that directly applies to gate-level netlists in an architecture-agnostic fashion like the pipeline-level reconfiguration, as will be discussed in Chap. 3.

## 2.8   Static Random Access Memory (SRAM)

SRAMs are fundamental building blocks of digital sub-systems, and are based on memory bitcells using positive feedback to store data in a volatile form. SRAMs are much denser than latch- and flip-flop-based memories (e.g., by 15–30×), and much faster than off-chip DRAM memories (e.g., access time commonly smaller by approximately two orders of magnitude). For these reasons, SRAMs are also used to implement register files in processors, other than multi-level caches and scratch pad memories.

A simplified SRAM floorplan is depicted in Fig. 2.21, where the row decoder selects one of the $N$ rows of each sub-bank and is shared between two horizontally adjacent sub-banks. The row decoder also includes the wordline drivers to speed up the charge/discharge of the input capacitance associated with the bitcells sharing the same wordline, and the wordline capacitance. The row decoder logic and wordline drivers are pitch-matched according to the SRAM bitcell height.

The horizontal middle portion in Fig. 2.21 consists of the column-wise circuitry for bitline voltage setting (e.g., precharge, write drivers) and read out (e.g., column multiplexer, sense amplifier). The center part of the memory (CTRL in Fig. 2.21) contains the control logic for timing generation, as pulses governing both read and write operation need to be delivered and sometimes tuned. In general, every sub-bank row contains $M$ words as in Figs. 2.21 and 2.22, each comprising $L$ bits. In each row, the first $M$ bits correspond to bit 0 of the $i$-th word with $i = 0 \ldots M - 1$. This pattern is repeated $L$ times to cover all the bits in a word. For each word bit position, one of the $M$ adjacent bits is selected via column multiplexing, as shown in Fig. 2.21. Column multiplexing permits to share the same sense amplifier among $M$ words in the same row, selecting only one of them at a time. In turn, this reduces

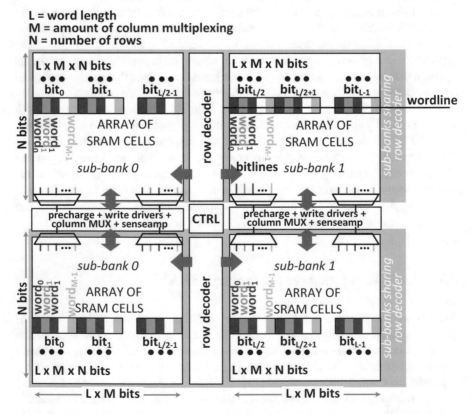

**Fig. 2.21** Floorplan of an SRAM memory with four sub-banks

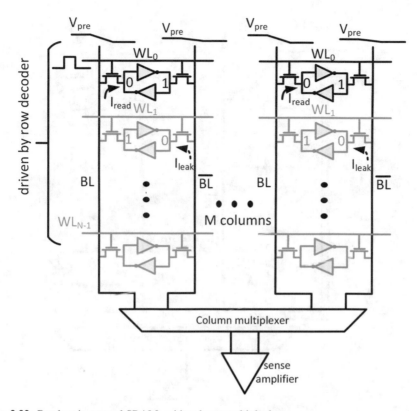

**Fig. 2.22** Readout in general SRAMs with column multiplexing

the overall sense amplifier area. From a layout viewpoint, this allows sense amplifier pitch matching with the bitcell, in spite of the fact that the width of the former is much larger than the latter.

A typical readout architecture is shown in Fig. 2.22. The row decoder activates only one wordline corresponding to the incoming row address. While reading, the bitlines are usually precharged to $V_{DD}$, and then their voltage starts falling at the rate of $I_{read}/C_{BL}$, where $I_{read}$ and $C_{BL}$ are the bitcell read current and the bitline capacitance, respectively. The differential voltage across the two complementary bitlines in a column becomes positive or negative, depending on which bitline is discharged, as determined by the content of the bitcell activated by the selected wordline. This small-signal change in the differential bitline voltage is then applied to the sense amplifier, after passing through the column MUX in Fig. 2.22. The sense amplifier resolves the bitline voltage to a full-swing output. The time taken to develop enough differential voltage for bitlines typically takes the largest fraction of the access time [31]. In other words, the read speed is strongly dependent on $I_{read}$.

Regarding the write operation, Fig. 2.23 shows the related operation, which is similar to read excepting from the fact that one of the two bitlines is discharged to ground, and the other is at $V_{DD}$. The rate of discharge of the bitline during write

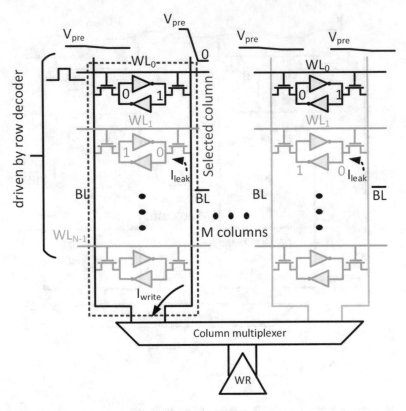

**Fig. 2.23** Write operation in column-multiplexed SRAM array

operation is given by $I_{\text{write}}/C_{\text{BL}}$, where $I_{\text{write}}$ is the bitline write current generated by the write drivers in Fig. 2.21. In most practical cases, $I_{\text{write}} >> I_{\text{read}}$ because $I_{\text{write}}$ is determined by the rather strong write driver gate, whereas and $I_{\text{read}}$ is only the on-current of the very small-sized access transistor of the SRAM bitcell. Therefore, the read speed of SRAM usually sets the SRAM speed bottleneck.

## 2.9   Methods for SRAM Speed-Up via Reconfigurable Array Organization

Reconfiguration in the SRAM is required for consistent memory-processor speed-up, when the processor core has a reconfigurable microarchitecture as discussed in the previous sections of this chapter.

In this section, the row aggregation technique to speed-up and dynamically adjust the access time is briefly discussed, with more details being provided in Sect.

3.10. To shorten the read access time, memory row aggregation doubles the effective six-transistor bitcell read current (i.e., bringing it to $2I_{read}$) being drawn from the bitline capacitance. In turn, this accelerates the bitline discharge by a factor of two, as shown in Fig. 3.18. To allow power-performance flexibility, row aggregation is selectively enabled to speed-up the read access time when needed (M+ mode), compared to the baseline array organization (M mode). Row aggregation is achieved through the simple modification in the row decoder in Fig. 3.19, as discussed in Sect. 3.10. This requires only a minor modification of an existing SRAM macro generated by a commercial memory compiler, as opposed to prior art that requires very substantial SRAM design effort (e.g., [32, 33]).

Being simultaneously activated, in the two active rows and the associated wordlines receive a wordline pulsewidth that is adjusted to shorten the allotted bitline discharge time. As discussed in Chap. 3, the adoption of row aggregation does not require any change in the software stack, as correct memory mapping simply requires compiler execution under the available address space.

## 2.10  Conclusion

In this chapter, microarchitecture reconfiguration has been introduced to extend the power-performance tradeoff beyond allowed by wide voltage scaling. The dynamically adaptable pipeline approach was shown to be adoptable in both application-specific accelerators and general-purpose processors, using the pipeline- and the thread-level reconfiguration. Finally, memory reconfiguration was briefly introduced to achieve coordinated power-performance scaling in both logic and memory. All these methods are minimally intrusive in terms of hardware and software designs, and hence represent a drop-in solution to make an existing microarchitecture reconfigurable.

## References

1. M. Alioto, E. Consoli, G. Palumbo, *Flip-Flop Design in Nanometer CMOS—from High Speed to Low Energy* (Springer, Berlin, 2015)
2. N. Weste, D. Harris, *CMOS VLSI Design*, 4th edn. (Pearson Education, London, 2011)
3. I. Sutherland, B. Sproull, D. Harris, *Logical Effort: Designing Fast CMOS Circuits* (Morgan-Kaufmann, San Francisco, CA, 1999)
4. J.L. Hennessy, D.A. Patterson, *Computer Architecture: A Quantitative Approach*, 6th edn. (Morgan Kaufmann, San Francisco, CA, 2019)
5. D. Markovic, R.W. Brodersen, *DSP Architecture Design Essentials* (Springer, Berlin, 2012)
6. K. Parhi, *VLSI Digital Signal Processing Systems: Design and Implementation* (Wiley, New York, 1999)
7. S. Jain, L. Lin, M. Alioto, Design-oriented energy models for wide voltage scaling down to the minimum energy point. IEEE Trans. CAS Pt. I **64**(12), 3115–3125 (2017)
8. M. Alioto (ed.), *Enabling the Internet of Things* (Springer, Berlin, 2017)

9. B.H. Calhoun, A.P. Chandrakasan, B.D. Aids, Characterizing and modeling minimum energy operation for subthreshold circuits, in *Proceedings of the International Symposium on Low Power Electronics and Design 2003, Newport Beach, CA*, (2004), pp. 90–95

10. T. Burd, T. Pering, A. Stratakos, R. Brodersen, A dynamic voltage scaled microprocessor system, in *IEEE ISSCC Digest of Technical Papers, San Francisco, CA*, (2000), pp. 294–295

11. S. Jain, S. Khare, S. Yada, V. Ambili, P. Salihundam, S. Ramani, S. Muthukumar, M. Srinivasan, A. Kumar, S. Kumar, R. Ramanarayanan, V. Erraguntla, J. Howard, S. Vangal, S. Dighe, G. Ruhl, P. Aseron, H. Wilson, N. Borkar, V. De, S. Borkar, A 280mV-to-1.2V wide-operating-range IA-32 Processor in 32nm CMOS, in *IEEE ISSCC Digest of Technical Papers, San Francisco, CA*, (2012), pp. 66–68

12. W. Wang, P. Mishra, System-wide leakage-aware energy minimization using dynamic voltage scaling and cache reconfiguration in multitasking systems. IEEE Trans. VLSI Syst. **20**(5), 902–910 (2012)

13. A. Chandrakasan, D. Daly, D. Finchelstein, J. Kwong, Y. Ramadass, M. Sinangil, V. Sze, N. Verma, Technologies for ultradynamic voltage scaling. Proc. IEEE **98**(2), 191–214 (2010)

14. M. Seok, D. Jeon, C. Chakrabati, D. Blaauw, D. Sylvester, Extending energy-saving voltage scaling in ultra low voltage integrated circuit designs, in *Proceedings of 2012 IEEE International Conference on IC Design & Technology, Austin, TX*, (2012), pp. 1–4

15. D. Jacquet, F. Hasbani, P. Flatresse, R. Wilson, F. Arnaud, G. Cesana, P. Magarshack, A 3 GHz dual core processor ARM Cortex TM-A9 in 28 nm UTBB FD-SOI CMOS with ultra-wide voltage range and energy efficiency optimization. IEEE J. Solid State Circuits **49**(4), 812–826 (2014)

16. F. Abouzeid, S. Clerc, B. Pelloux-Prayer, F. Argoud, P. Roche, 28nm CMOS, energy efficient and variability tolerant, 350mV-to-1.0V, 10MHz/700MHz, 252bits frame error-decoder, in *Procdings of ESSCIRC 2012, Bordeaux, France*, (2012), pp. 153–156

17. S. Hsu, A. Agarwal, M. Anders, S. Mathew, H. Kaul, F. Sheikh, R. Krishnamurthy, A 280mV-to-1.1V 256b reconfigurable SIMD vector permutation engine with 2-dimensional shuffle in 22nm CMOS, in *IEEE ISSCC Digest of Technical Papers, San Francisco, CA*, (2012)

18. Y. Zhang et al., 8.8 iRazor: 3-transistor current-based error detection and correction in an ARM Cortex-R4 processor, in *IEEE ISSCC Digest of Technical Papers, San Francisco, CA*, (2016), pp. 160–162

19. S. Hanson, B. Zhai, K. Bernstein, D. Blaauw, A. Bryant, L. Chang, K.K. Das, W. Haensch, E.J. Nowak, D.M. Sylvester, Ultralow-voltage, minimum-energy CMOS. IBM J. Res. Dev. **50**(4/5) (2006)

20. H. Shimada, H. Ando, T. Shimada, Pipeline stage unification: a low-energy consumption technique for future mobile processors. Proc. Int. Sympos. Low Power Electr. Design **2003**, 326–329 (2003)

21. A. Efthymiou, J.D. Garside, Adaptive pipeline depth control for processor power-management, in *Proceedings of IEEE International Conference on Computer Design: VLSI in Computers and Processors*, (2002), pp. 454–457

22. S. Chellappa, C. Ramamurthy, V. Vashishtha, L.T. Clark, Advanced encryption system with dynamic pipeline reconfiguration for minimum energy operation, in *Proceedings of 16th International Symposium on Quality Electronic Design (ISQED), Santa Clara, CA*, (2015), pp. 201–206

23. S. Vijayalakshmi, A. Anpalagan, I. Woungang, D.P. Kothari, Power management in multi-core processors using automatic dynamic pipeline stage unification, in *2013 International Symposium on Performance Evaluation of Computer and Telecommunication Systems (SPECTS), Toronto (Canada)*, (2013), pp. 120–127

24. H. Jacobson, Improved clock-gating through transparent pipelining, in *Proceedings of the International Symposium on Low Power Electronics and Design 2004, Newport Beach, CA*, (2004), pp. 26–31

25. S. Manne, A. Klauser, D. Grunwald, Pipeline gating: speculation control for energy reduction, in *Proceedings of 25th Annual International Symposium on Computer Architecture, Barcelona, Spain*, (1998), pp. 132–141

26. H. Shimada, H. Ando, T. Shimada, A hybrid power reduction scheme using pipeline stage unification and dynamic voltage scaling, in *Proceedings of IEEE COOL Chips*, (2006), pp. 201–214
27. L. Lin, S. Jain, M. Alioto, Integrated power management and microcontroller for ultra-wide power adaptation down to nW, in *2019 Symposium on VLSI Circuits, Kyoto, Japan*, (2019), pp. C178–C179
28. L. Lin, S. Jain, M. Alioto, A 595pW 14pJ/Cycle microcontroller with dual-mode standard cells and self-startup for battery-indifferent distributed sensing, in *2018 IEEE International Solid-State Circuits Conference—(ISSCC), San Francisco, CA*, (2018), pp. 44–46
29. D. Rossi, A. Pullini, I. Loi, M. Gautschi, K. Frank, A.T. Gurkaynak, et al., Energy-efficient near-threshold parallel computing: the PULPv2 cluster. IEEE Micro. **37**(5), 20–31 (2017)
30. N. Weaver, Y. Markovskiy, Y. Patel, J. Wawrzynek, Post-placement C-slow retiming for the Xilinx Virtex FPGA. FPGA (2003)
31. S. Jain, L. Lin, M. Alioto, Drop-in energy-performance range extension in microcontrollers beyond VDD scaling, in *2019 IEEE Asian Solid-State Circuits Conference, Macau*, (2019), pp. 125–128
32. H. Fujiwara, S. Okumura, Y. Iguchi, H. Noguchi, H. Kawaguchi, M. Yoshimoto, A 7T/14T dependable SRAM and its array structure to avoid half selection, in *22nd International Conference on VLSI Design*, (2009), pp. 295–300
33. S. Okumura, S. Yoshimoto, K. Yamaguchi, Y. Nakata, H. Kawaguchi, M. Yoshimoto, 7T SRAM enabling low-energy simultaneous block copy. CICC (2010)

# Chapter 3
# Automated Design Flows and Run-Time Optimization for Reconfigurable Microarchitecures

**Abstract** In this chapter, a systematic methodology is introduced to design reconfigurable microarchitectures through automated and architecture-agnostic design flows. The main goal is to enrich a baseline microarchitecture with additional registers for throughput enhancement and then make selected registers bypassable to flexibly switch among different microarchitectures. Similarly, design methodologies for reconfigurable SRAM memories are described. As common thread, drop-in solutions for existing architectures allowing the above capability at very low design effort are discussed.

**Keywords** Design methodology · Digital design · Design flow · Gate-level netlist manipulation · Automated synthesis and place&route · CAD algorithm · Reconfigurable microarchitecture · Reconfigurable SRAM · Pipestage-level reconfiguration · Thread-level reconfiguration · Bank-level reconfiguration · Pipestage · Pipeline stage · Fan-out-of-4 delay · Re-pipelining · Register level · Linear pipeline · Feedforward pipeline · Feedback pipeline · Loop · Register branch · Dynamic energy · Leakage energy · Above-threshold region · Near-threshold region · Sub-threshold region · Leakage-dynamic energy ratio · Fixed microarchitectures · Minimum energy point (MEP) · Dynamically adaptable pipelines · Dynamic voltage frequency scaling · Power mode · Bypassable register · Bypassable flip-flop · Non-bypassable register · Non-bypassable flip-flop · EDA tool · Retiming · Delay overhead · Register bypassing · Flip-flop bypassing · Throughput enhancement · Control flow · Pipeline bubble · Time-interleaved microarchitecture · Time interleaving · Input stream · Instruction stream · Channel · Gate-level netlist · SRAM · Instruction memory · Data memory · Column multiplexing · Column multiplexer · Bitline · Wordline · Sense amplifier · Precharge driver · Write driver · Bitcell · Memory bank · Memory sub-bank · Reconfigurable array organization · Access time · Row aggregation · Drop-in microarchitecture reconfiguration · Electronic design automation (EDA) · Pipeline stage unification · AES · Transparent register · Static microarchitecture · Dynamic microarchitecture · Cycle-level timing · Netlist · Skeleton graph · Register identification · Bypassable register replacement · Netlist-to-skeleton graph · Graph weighting · Level identification · Cutset · Feedforward cutset · Cutset identification · Cutset-to-pipeline mapping · Even-numbered register identification · Script · Place&route (PNR) · Behavioral RTL · Register transfer level (RTL) · Weighted skeleton graph · Tcl script · Cutset-based identification · Non-linear pipelines ·

© Springer Nature Switzerland AG 2020
S. Jain et al., *Adaptive Digital Circuits for Power-Performance Range beyond Wide Voltage Scaling*, https://doi.org/10.1007/978-3-030-38796-9_3

Linear pipelines · Register insertion · Register merging · Reconvergent path · Branching path · Graph · Netlist graph · Hash table · Flip-flop reset · Graph edge · Graph node · Dummy node · Dummy edge · Static timing analysis (STA) · Weight · Graph traversal · Depth-first traversal · Row decoder · Reconfigurable decoder

In this chapter, a systematic methodology is introduced to design reconfigurable microarchitectures through automated and architecture-agnostic design flows. The main goal is to enrich a baseline microarchitecture with additional registers for throughput enhancement and then make selected registers bypassable to flexibly switch among different microarchitectures. Similarly, design methodologies for reconfigurable SRAM memories are described. As common thread, drop-in solutions for existing architectures allowing the above capability at very low design effort are discussed.

To suppress the need for *ad hoc* arrangements and assure general applicability, the design flows in this chapter solely rely on commercial electronic design automation (EDA) tools and appropriate scripts to integrate them into an automated design flow. All scripts have been made publicly available, as detailed in the Appendix. The resulting design methodology is shown to be architecture-agnostic and can hence be applied to third-party intellectual properties (IPs), in addition to proprietary designs.

## 3.1    Prior Art in Reconfigurable Microarchitectures

The energy consumption of systems on chip (SoCs) is well known to have gained equal importance as performance, due to its impact on the battery life in self-powered systems, and on performance in high-performance systems [1]. Indeed, SoCs typically operate in a power limited regime, ranging from server to mobile and IoT applications [2]. To save energy under a wide range of workloads and performance targets, wide voltage scaling has been extensively adopted in the last two decades [3–10]. Indeed, dynamic voltage scaling (DVS) reduces the energy quadratically at above-threshold voltages. On the other hand, DVS loses its edge when the supply voltage of an SoC designed for above-threshold voltages is down-scaled close to the near-threshold or the sub-threshold region [11–14]. This occurs because the energy reaches a minimum energy point (MEP), which lies in either of these regions in practical cases.

In general, energy efficiency at any given supply voltage is determined by the adopted microarchitecture, which has to be properly chosen to optimally balance the dynamic, clocking, and leakage energy contributions [15–17]. In prior art, co-optimization of a static microarchitecture and the supply voltage has been widely explored to reduce the energy for a given voltage or a throughput target [18].

Unfortunately, the adoption of a static microarchitecture is able to improve the energy efficiency only in a narrow range of voltages or performance targets. This is due to the strong dependence of the dynamic-to-leakage energy ratio on the supply voltage and the throughput target [19], which makes an energy-optimal microarchitecture at a given voltage less optimal at a significantly different voltage.

As further prior art, reconfigurable pipelines were investigated in [20–25], although for purposes that are fundamentally different from the objectives pursued in this book. For example, [20–22] proposed pipeline stage unification as a way to reduce energy consumption at the cost of lower performance at a given supply voltage (i.e., no voltage scaling is allowed), as an alternative option to the energy-performance tradeoff offered by voltage scaling. Also, [23] demonstrated pipeline stage unification in an AES core for a narrow range of above-threshold voltages. The work in [24] explored the benefits of selectively transparent registers to reduce the clocking energy at the pipestage level without any performance loss, keeping the supply voltage constant. Pipeline stage unification has also been explored in [25] to tradeoff hazards (i.e., pipeline stalls) and performance in microprocessor architectures that are explicitly conceived to incorporate microarchitecture reconfiguration, with the supply voltage being kept constant. This offers an alternative knob to tradeoff energy and performance at a given voltage and cannot be applied to the existing microarchitectures. Overall, prior art in reconfigurable microarchitectures at the pipestage and thread level is restricted to a fixed supply or a narrow voltage scaling and does not explore the opportunities offered by joint microarchitecture reconfiguration and voltage scaling. In addition, previous microarchitecture reconfiguration techniques have limited applicability since they are based on *ad hoc* microarchitectural rearrangements and design methodologies.

Recently, dynamic microarchitecture reconfiguration with granularity down to the pipeline stage (or thread) level has been introduced to optimally balance the above energy contributions at different voltages, when the supply voltage is dynamically scaled in a wide range [26–28]. In general, reconfigurable microarchitectures adjust the dynamic-to-leakage energy ratio by adapting the logic depth to either the targeted voltage or the targeted throughput. Higher logic depths are obtained by dynamically and selectively bypassing registers, which in turn reduces the clocking energy and increases the leakage energy due to the increased clock cycle at a given voltage [26–28]. Conversely, lower logic depths are achieved by keeping all registers active (i.e., without bypassing any), which increases the clocking energy and reduces the leakage energy. Accordingly, joint logic depth adjustment and voltage scaling permits to dynamically compensate the imbalance among the above energy contributions due to significant voltage changes compared to a fixed design point. In turn, this allows to select the most energy-efficient microarchitecture among the available configurations to maintain nearly energy optimality or improve performance beyond allowed by conventional voltage scaling with a static microarchitecture [26–28].

## 3.2   Overview of Systematic Methodologies and Design Flows for Microarchitectural Reconfiguration

In this chapter, a unified design methodology to systematically and automatically design reconfigurable microarchitectures is presented, starting from a fixed baseline microarchitecture (e.g., an existing proprietary or third-party design). Guidelines are provided to quantify the cost of reconfiguration before actually carrying out the design. The design flow is based on proper algorithms that allow to cluster individual flip-flops into the registers that belong to each pipestage and to identify which registers should be made bypassable to provide the best energy efficiency. The design methodology described in the following empowers traditional wide voltage–frequency scaling with the ability to dynamically and simultaneously co-optimize the logic depth of the microarchitecture.

The overall design flow for pipeline- and thread-level reconfiguration is respectively shown in Fig. 3.1a, b. The flow in Fig. 3.1a (Fig. 3.1b) is applicable to any generic microarchitecture of application-specific accelerators (general-purpose processor). These figures show that the general flow is the same, with the thread-level reconfiguration flow in Fig. 3.1b being simpler than the pipeline-level one, as discussed in Sect. 3.4. Overall, the unitary flow in Fig. 3.1a, b generalizes the design methodologies for pipeline- and thread-level reconfiguration in [26–28].

The design flow in Fig. 3.1a, b comprises four basic reconfiguration steps, which are detailed in Sect. 3.3 for pipeline-level reconfiguration and Sect. 3.4 for thread-level reconfiguration. In summary, Step 1 modifies the netlist of the baseline static microarchitecture, introducing an appropriate number of conventional registers to

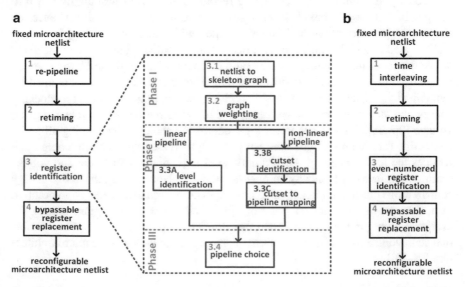

**Fig. 3.1** Unitary design flow to translate a baseline fixed microarchitecture into (**a**) reconfigurable one at the pipeline level [27] and (**b**) reconfigurable one at the thread level [28]

create the deepest pipeline configuration to achieve the maximum throughput target. Such register insertion is guaranteed to preserve functionality and cycle-level timing, although it might affect the latency. For pipeline-level reconfiguration, this step can be skipped if the baseline architecture already meets the throughput specification. For thread-level reconfiguration, the insertion of registers in Step 1 is invariably needed to enable time-interleaved operation, as detailed in Sect. 3.4.

In Step 2, the re-pipelined (time-interleaved) microarchitecture is retimed to balance the pipestage delays to maximize the operating frequency and hence the throughput. At the end of Step 2, the resulting design is faster than the baseline design, but still lacks reconfiguration. Step 3 identifies the registers that offer the best energy efficiency improvement, while maintaining the original functionality as the baseline microarchitecture and maintaining balanced pipestages across different configurations. In Step 4, the registers that were found to be the best candidates for bypassing are replaced with the bypassable version, thus resulting into the desired reconfigurable microarchitecture.

The microarchitectural pipestage reconfiguration flow in Fig. 3.1a, b solely relies on commercial EDA tools, as well as custom Tcl scripts to manipulate the netlists at various steps and assure smooth tool integration. Figure 3.2 shows the resulting design flow for non-linear pipelines and the integration of custom scripts and EDA tools. Very similar derivations hold for the simpler case of linear pipelines, which is omitted accordingly.

## 3.3 Automated Design Flow for Pipeline-Level Reconfiguration: Re-pipelining and Retiming (Steps 1–2)

Steps 1 and 2 in Fig. 3.1a are discussed in Sects. 3.3.1 and 3.3.2, respectively.

### 3.3.1 Re-pipelining (Step 1)

In a static microarchitecture under wide voltage scaling, clock frequency and throughput are upper bounded by their value at the nominal voltage $V_{DD, nom}$, which is set by the resulting logic depth and the adopted technology. Before the baseline microarchitecture is made reconfigurable, it has first to meet the specified maximum clock frequency $f_{CK, max}$. If the latter is not met, the pipestage logic depth (i.e., the clock cycle) is routinely reduced through re-pipelining.

Re-pipelining (Step 1 in Fig. 3.1a) consists of the proper insertion of additional registers to slice the combinational logic into a larger number of shorter pipestages, while preserving the overall functionality. The larger number of registers in paths

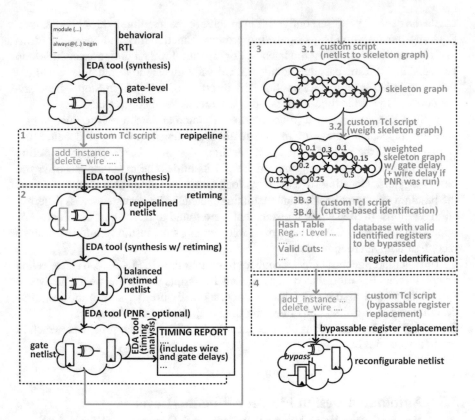

**Fig. 3.2** Detailed design flow for non-linear pipelines from Fig. 3.1a, and integration of EDA tools and custom scripts. The design flow for linear pipelines is very similar and hence omitted (only the steps in Phase II in Fig. 3.1a are modified)

from inputs to outputs increases the latency, while improving the throughput[1] [18]. As common design practice, the number of registers to be inserted during re-pipelining is found by progressively inserting more registers and retiming, until the clock frequency target $f_{CK, max}$ is achieved. The clock frequency is evaluated through post-synthesis (or post-place&route) timing analysis through commercial EDA tools (e.g., [29, 30]). Register insertion is performed at either the RTL level (e.g., Verilog description, if available) or the netlist level in the original static microarchitecture (e.g., gate-level netlist provided by a third-party IP vendor or generated through synthesis from the RTL design if available). For simplicity, registers are often inserted from the inputs or the outputs as in Fig. 3.3 and then pushed into the core logic through retiming as discussed in the next subsection [31].

---

[1] It is well known that re-pipelining does not provide any throughput improvement, if the critical path is in a feedback loop [32]. This is an intrinsic limitation of any pipelined microarchitectures, either static or reconfigurable.

**Fig. 3.3** Register insertion from inputs and outputs for re-pipelining (Step 1)

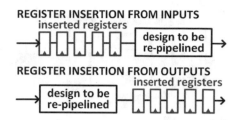

## 3.3.2   Retiming (Step 2)

At Step 2 in Fig. 3.1a, the design including the registers added at Step 1 is then retimed to move the registers inside the combinational logic (i.e., pushing the added registers in Fig. 3.3) and balance out the pipestage delays to minimize the clock frequency. Automatic retiming is supported by all major commercial EDA tools, and its effectiveness in terms of clock frequency improvement depends on the specific topology of the microarchitecture (i.e., type of elementary microarchitectures employed in it).

With reference to the elementary microarchitectures summarized in Chap. 2, retiming maintains the same register count as the added registers are simply redistributed across the combinational logic, as there are no branches as summarized in Fig. 3.4a. In the example presented in this figure, the number of registers before and after retiming is equal to three, as register A is simply pushed forward without creating any new register. On the other hand, feedforward microarchitectures inherently have branching paths, and a register moved from before to after a branch translates into a larger number of registers equal to the number of branches as in Fig. 3.4b. In the example in this figure, the register B in gray is moved downstream after the three branches and is replaced by three registers B1, B2, and B3. Hence, feedforward microarchitectures generally suffer from a larger register overhead when being re-pipelined into a deeper structure, compared to linear pipelines. Conversely, the presence of reconvergent paths in feedback elementary microarchitectures in Fig. 3.4c reduces the register count at the point of path reconvergence. Indeed, the registers originally belonging to the reconvergent paths are moved downstream and are hence merged into a single register in the subsequent path. In the example in Fig. 3.4c, the registers C1 and C2 before the reconvergence point are moved downstream, and merge into the single register C.

Once the baseline microarchitecture is enriched with an adequate number of additional registers and the resulting clock cycle is made small enough, the maximum clock frequency target is achieved at the end of Step 2 in Figs. 3.1a and 3.2. As will be required in the subsequent "graph weighting" (Step 3.1 in Fig. 3.1a), static timing analysis will be performed on the retimed design. To accurately account for the impact of wires, the retimed netlist is placed and routed (PNR) as depicted at the end of Step 2 in Fig. 3.2, and the post-PNR timing report is then extracted for later use in Step 3.2. If wires are known to have a minor effect for a given microarchitecture (e.g., there are no high-fanout nodes), the retimed post-synthesis netlist

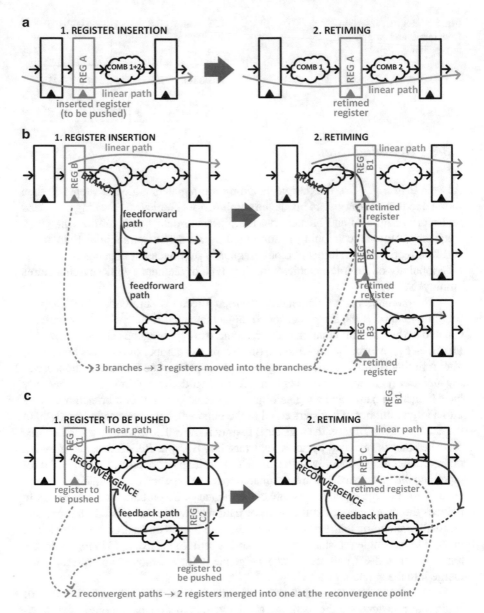

**Fig. 3.4** Register count in input-output paths in (**a**) linear, (**b**) feedforward (i.e., with branching paths), and (**c**) feedback pipelines (i.e., having reconvergent paths). From these figures, register insertion and retiming (i.e., re-pipelining) maintain the same register count as in the original microarchitecture in linear pipelines, whereas it increases (decreases) in feedforward (feedback) elementary microarchitectures

can be used directly in Step 3.2 to reduce the design effort. Reconfiguration is then introduced through Steps 3–4 in Fig. 3.1a, as discussed in the following sections.

## 3.4 Automated Design Flow for Pipeline-Level Reconfiguration: Register Identification (Step 3, Phase I)

### 3.4.1 Netlist to Skeleton Graph (Step 3.1, Phase I)

The "register identification" step is the first phase of Step 3 in Fig. 3.1a and aims to identify the registers that can be made bypassable, while preserving the original functionality and assuring balanced pipestages in all configurations (i.e., with registers in either bypass or normal operation). Formally, register bypassing is equivalent to register removal for re-pipelining into a shallower pipeline [32]. Hence, the gate-level netlist must be properly manipulated to maintain the same relative timing of signals at the input of each and all pipestages and hence correct overall functionality [31].

Phase I of register identification consists of two sub-steps, i.e., skeleton graph conversion and skeleton graph weighting. The first sub-step converts the gate-level netlist generated at Step 2 (Sect. 3.3.2) into a "skeleton" graph that contains only its registers, while suppressing any combinational logic. This permits to drastically simplify the graph associated with the microarchitecture being designed, leveraging the fact that bypassable register identification and the general pipeline-level abstraction are not affected by the specific combinational logic in each pipestage. In other words, the skeleton graph still retains the necessary information on the pipeline structure (e.g., how elementary pipelines are combined).

In detail, the gate-level netlist generated at Step 2 is first parsed and converted to a *skeleton graph*. This is achieved by first creating the *netlist graph*, which converts each input-output pair in every logic gate into a graph edge, and each wire (i.e., the output of each logic gate) into a graph node. By construction, the edges associated with all input-output pairs of each logic gate converge into the graph node associated with the wire at the logic gate output. For example, the XOR gate in Fig. 3.5 with two inputs ($In_1$, $In_2$) and output wire $w_1$ are converted into two skeleton graph edges $\langle In_1, w_1 \rangle$, and $\langle In_2, w_2 \rangle$ converging into the node $w_1$, as shown at Step 3.1.1 in Fig. 3.5. From a data structure viewpoint, the output of Step 3.1.1 in Fig. 3.5 is stored in the form of a hash table storing a *key* and the corresponding *value*. The *key* in the hash table represents a graph edge (or the combination of two graph nodes being connected by an edge). The *value* of the corresponding key contains the necessary information on the gate type (e.g., AND, OR) and the ports of the gates (e.g., inputs, flip-flop reset ports) along with the edges that they are connected to, as reported in the original instance of the gate within the original netlist.

The resulting *netlist graph* from Step 3.1 is converted to the *skeleton graph* by replacing any node associated with combinational gates with an edge, as shown in

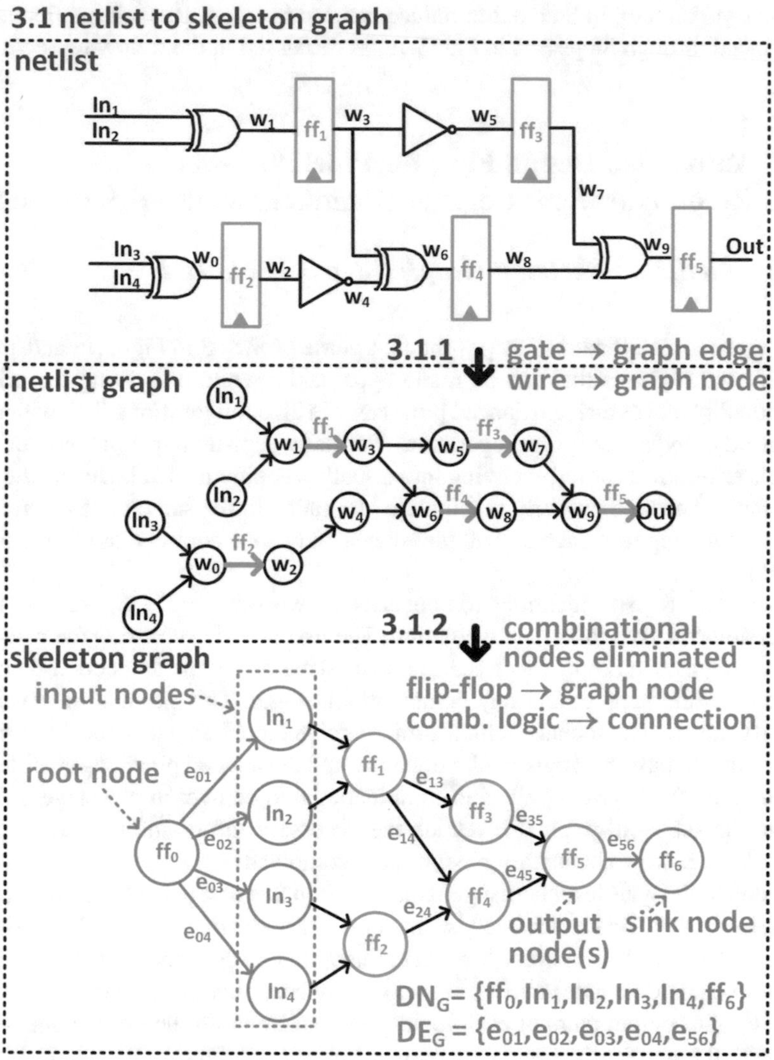

**Fig. 3.5** Conversion of a gate-level netlist to a skeleton graph (dummy nodes and edges in gray, skeleton nodes associated with flip-flops are in blue)

Step 3.1.2 in Fig. 3.5. The skeleton graph contains just enough information about the sequential paths and hence the microarchitecture topology, while it ignores the combinational logic to substantially reduce the size of the graph database. From a graph point of view, subsequent edges associated with combinational gates are lumped into a single edge, as shown in the example in Fig. 3.5. The skeleton graph is obtained from the netlist graph through depth-first traversal from primary inputs to the outputs of the netlist graph. To keep the information about primary inputs intact in spite of the subsequent edge lumping, each input port is converted into a node, and the gates driving the output ports are also converted to graph nodes. The resulting skeleton graph hence contains only the basic information on how signals flow between flip-flops. In Fig. 3.5, the source, sink, and input nodes are always categorized as dummy (in gray color), since they do not represent actual flip-flops. On the other hand, output nodes are categorized as dummy only if the output port is being driven by a combinational gate. Instead, if an output is driven by a flip-flop, the output port is associated with a skeleton node.

Hereafter, the set of dummy nodes of a skeleton graph $G$ will be represented as $DN_G$. Similarly, the edges connecting dummy nodes or the edges connecting the output nodes to the sink node are categorized as dummy edges, whose set is represented as $DE_G$. For example, in Fig. 3.5 the edge $e_{01}$ connecting the dummy nodes $ff_0$ and $In_1$ is a dummy edge, and the same holds for the edges $e_{02}, e_{03}, e_{03}, e_{04},$ and $e_{56}$ as there is no combination logic represented by these edges.

### 3.4.2 Weighted Skeleton Graph (Step 3.2, Phase I)

Step 3.2 aims to back-annotate the skeleton graph with the delay of the most critical combinational path encountered when traversing the graph from any launching flip-flop to any subsequent capturing flip-flop (e.g., the delay of the combinational path $e_{13}$ from flip-flop $ff_1$ to $ff_3$ in Fig. 3.5). This timing information will be crucial at the bypassable register choice at Step 3.4 in Fig. 3.1a, which clearly needs to be timing-aware.

The back-annotation process is exemplified in Fig. 3.6, where the skeleton graph being back-annotated is taken from Fig. 3.5. This is achieved by automatically converting the skeleton graph into a timing generation script that is readable by the timing analysis tool at the end of Step 3.2 in Figs. 3.1a and 3.2. Once the latter generates the timing report with the desired delay information, the timing report is parsed to extract each critical path delay, which is in turn dumped back into the skeleton graph to update the graph database. By construction, the resulting weighted

**Fig. 3.6** The weighted skeleton graph (Step 3.2) is derived by enriching the skeleton graph in Fig. 3.5 with edge weights equal to the critical path delay between pairs of adjacent flip-flops (i.e., graph nodes), as evaluated through the timing analysis tool

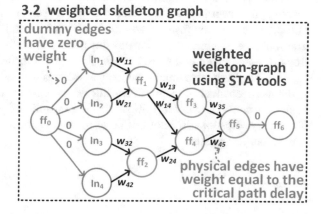

skeleton graph in Fig. 3.6 has zero weight in dummy edges, and non-zero weight in physical edges as provided by the reports generated by static timing analysis tool.

## 3.5  Automated Design Flow for Pipeline-Level Reconfiguration: Register Identification in Linear Pipelines (Step 3, Phase II)

In this section, Phase II of register identification are discussed for linear pipelines, as relevant to Step 3.3A in Fig. 3.1a ("level identification"). In this step, flip-flops are clustered into the corresponding register level they belong to, as this will be necessary to identify the registers to be bypassed later on. This is needed because the retimed gate-level netlist from Step 2 is flattened, and hence loses the relationship between flip-flops and the register they belong to.

In linear pipelines, flip-flop clustering into pipeline registers is achieved through recursive traversal of the skeleton graph obtained at Step 3.1 (Sect. 3.4.1). Pipeline registers are numbered according to the register level they occupy, as defined by the number of registers encountered from the primary inputs to a given pipestage. For example, in the linear pipeline at the top of Fig. 3.5, $ff_1$ and $ff_2$ belong to the register level 1 (i.e., only one register is crossed in the path from the primary inputs to the outputs of $ff_1$ and $ff_2$), $ff_3$ and $ff_4$ are associated with the register level 2, whereas $ff_5$ belongs to the register level 3.

By definition, the register level of a register depends on the specific path that is considered to reach it, starting from the primary inputs. However, in linear pipelines, the register level of any register is independent of the considered path (see Chap. 2 and Fig. 3.4a). This is because the registers are visited always in the same order for any path from the primary inputs to any intermediate signal, by definition of linear pipeline. In other words, the register level of any register in a linear pipeline is a property of the register itself, rather than depending on the considered path.

By definition, the register level of a given flip-flop is simply evaluated by progressively counting the number of the flip-flops that were previously crossed before reaching it. Accordingly, the skeleton graph is recursively traversed from input to output in a depth-first fashion, as shown in Step 3.3A in Fig. 3.7. For every skeleton graph node encountered (i.e., a flip-flop), the cumulative number of crossed flip-flops is incremented and stored in a hash table via recursive function calling as in Fig. 3.7. The hash table unambiguously associates the register level to each flip-flop, given its independence from the considered path. At the end of graph traversal, the hash table contains every flip-flop in the design, and its corresponding register level. Ultimately, flip-flops in the hash table belonging to the same level belong to the same pipeline register.

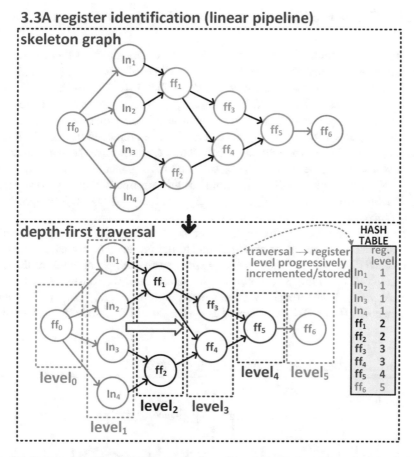

**Fig. 3.7** Register identification in linear pipelines through depth-first traversal of the skeleton graph (Step 3.3A in Fig. 3.1a), as applied to the skeleton graph derived in Fig. 3.6

## 3.6 Automated Design Flow for Pipeline-Level Reconfiguration: Register Identification in Non-linear Pipelines (Step 3, Phase II)

In this section, Phase II of register identification is discussed for non-linear pipelines. In microarchitectures that are not strictly linear, associating a flip-flop to a register level is less obvious than in linear ones. Indeed, a flip-flop in a non-linear microarchitecture may be associated with one level if accessed through a path or another if accessed through a different one. This makes register identification (Step 3 in Fig. 3.1a) more difficult than linear pipelines, as addressed in the remainder of this section.

In non-linear pipelines, Phase II involves two steps: the "cutset identification" and the "cutset-to-pipeline mapping" (Step 3.3B and 3.3C in Fig. 3.1a, respectively). These steps are respectively described in Sects. 3.6.2 and 3.6.3, after introducing the definition and some properties of graph cutsets in Sect. 3.6.1.

### 3.6.1   *Graph Feedforward Cutsets and Properties*

Register identification in non-linear pipelines leverages the well-known property that removing registers (or bypassing, in the context of this work) belonging to a feedforward cutset provably preserves functionality [31, 32]. To this aim, let us consider a directed graph $G = (N, E)$, as defined by its set of nodes $N$ and edges $E$. In this graph, a cutset $c_i$ is defined as a minimal set of edges that yields two disjoint upstream and downstream sub-graphs $G_u$ and $G_d$, when removed from $G$. A cutset $c_i$ is of feedforward type when all edges connecting $G_u$ to $G_d$ have the same direction, which is assumed to be from $G_u$ to $G_d$ with no loss of generality. The set of nodes $N_{u,c_i}$ ($N_{d,c_i}$) belonging to $G_u$ ($G_d$) comprises the "upstream nodes" ("downstream nodes"). Similarly, the set of edges $E_{u,c_i}$ ($E_{d,c_i}$) belonging to $G_u$ ($G_d$) comprises the "upstream edges" ("downstream edges").

The above definitions are exemplified for the *skeleton graph* $G$ in Fig. 3.8, which explicitly includes non-linear pipelines due to the presence of branches departing from $\text{ff}_5$ and $\text{ff}_8$. Figure 3.9a shows $c_i = \{e_{23}, e_{54}\}$ as an example of cutset of the skeleton graph $G$ in Fig. 3.8, splitting the latter into the upstream sub-graph $G_u = \left(N_{u,C_i}, E_{u,C_i}\right)$ and the downstream sub-graph $G_d = \left(N_{d,C_i}, E_{d,C_i}\right)$. The edge between nodes $\text{ff}_i$ and $\text{ff}_j$ is represented as $e_{ij}$. The $k$th child of node $\text{ff}_i$ is represented

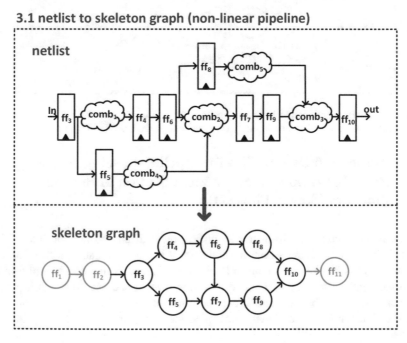

**Fig. 3.8** Example of gate-level netlist with non-linear pipeline and skeleton graph derivation. This example is used in the remainder of the section on non-linear pipelines, as the original gate-level netlist example in Fig. 3.5 actually referred to a simpler linear pipeline

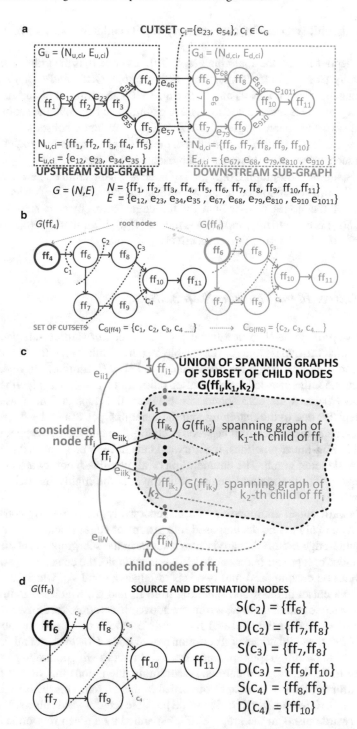

**Fig. 3.9** Examples of (**a**) upstream and downstream sub-graphs defined by a cutset, (**b**) spanning graph $G(n_i)$ of $n_i$, (**c**) spanning sub-graph $G(n_i, k_1, k_2)$ of graph $G(n_i)$ via the child nodes $k_1 \ldots k_2$, (**d**) source node $S(c_i)$, and destination node $D(c_i)$ associated with a cutset $c_i$

as $\text{ff}_{ik}$, and the child edges of $\text{ff}_i$ connecting it with its children nodes are represented as $e_{iik}$.

As further definition, the spanning graph $G(\text{ff}_i)$ of node $\text{ff}_i$ is the graph that can be spanned fully from node $\text{ff}_i$ downstream. For example, $G(\text{ff}_4)$ and $G(\text{ff}_6)$ represent the spanning graphs of nodes $\text{ff}_4$ and $\text{ff}_6$ in Fig. 3.9b. The entire set of cutsets associated with a graph $G$ is represented as $C_G$. For example, in Fig. 3.9b the set of cutsets associated with the spanning graph $G(\text{ff}_4)$ is $C_{G(\text{ff}_4)}$. When a subset $k_1...k_2$ of the child nodes $1...N$ of a node $n_i$ is considered, the union of all spanning graphs associated with such subset of child nodes is represented as $G(\text{ff}_i, k_1, k_2)_{ik}$. Fig. 3.9c exemplifies the spanning sub-graph $G(n_i, k_1, k_2)$ of node $n_i$ via the child edges from $e_{iik_1}$ to $e_{iik_2}$. Finally, the set of source nodes $S(c_i)$ (destination nodes $D(c_i)$) of a cutset $c_i$ comprises all the nodes from which all the edges of a cutset $c_i$ start (end). In Fig. 3.9d, $S(c_2)$ is the set $\{\text{ff}_6\}$, and $S(c_3)$ is the set $\{\text{ff}_7, \text{ff}_8\}$. Similarly, $D(c_2)$ and $D(c_3)$ are the sets $\{\text{ff}_7, \text{ff}_8\}$ and $\{\text{ff}_9, \text{ff}_{10}\}$, respectively.

## 3.6.2   Cutset Identification (Step 3.3B)

As a first step of Phase II in non-linear pipelines, feedforward cutsets need to be identified since the bypassable (or removable, in a fixed microarchitecture) pipeline registers certainly lie along a feedforward cutset [31, 32]. Similarly to linear pipelines (Sect. 3.5), the gate-level netlist after retiming of non-linear pipelines (Step 2 in Fig. 3.1a) loses the correspondence between flip-flops and their associated pipeline register due to the flattening and the renaming performed by the retiming tool. To re-create the association between each flip-flop and the corresponding register level in non-linear pipelines, feedforward cuts have to be identified in the corresponding skeleton graph. The enumeration of all possible feedforward cutset can be performed with the method below, which avoids the highly inefficient brute-force search.

Feedforward cutset enumeration can be recursively performed by considering that a cutset can always be decomposed into groups of edges (including the simple case of single-edge group) that are in turn a cutset in the sub-graphs spanning from all child nodes of a proper parent node [33]. For example, the cutset $c_i = \{e_{23}, e_{54}\}$ in Fig. 3.9a can be decomposed into two sets of edges $e_{23}$ and $e_{54}$, which are readily found to be a cutset in the graph spanning from $\text{ff}_4$ and $\text{ff}_5$, which are in turn child nodes of the same parent node $\text{ff}_3$. From the above observation, feedforward cutsets need to be searched in the space of the possible unions of cutsets in sub-graphs spanning from the child nodes of a common parent node. However, not all unions of the cutsets of such sub-graphs spanning from child nodes with common parents are guaranteed to be feedforward cutsets. Accordingly, the unions that do not form an actual feedforward cutset need to be discarded in this search. This leads to the algorithm in Fig. 3.10 for the identification of all possible feedforward cutsets. Basically, all feedforward cutsets in the graph $G(\text{ff}_{\text{root}})$ spanned by a given root node $\text{ff}_{\text{root}}$ are

---

**Algorithm: cutset identification**

---

**global sklgraph** $G(ff_{root})$ = *CreateGraph*(netlist)
**cutset** $C_{G(ffroot)}[]$ = *FindSetOfCutset*($ff_{root}$)

**cutset[]** *FindSetOfCutset* ($ff_{curr}$)
   **cutset** $C_{G(ffcurr)}[]$ = []
   **node** $ff_{children}[]$ = *GetChildren*($ff_{curr}$)
   **for** k = 1: ($ff_{children}$.*length*)
      **node** $ff_{child}$ = $ff_{children}[k]$
      **edge** $e_{curr}$ = {$ff_{curr}$,$ff_{child}$}
      **cutset** $C_{ffchild}$ = {$e_{curr}$}                                         $C_{G(ffcurr,k,k)}$
      $C_{G(ffcurr)}$ = *Merge*($C_{G(ffcurr,1,(k-1))}$, {*FindSetOfCutset*($ff_{child}$) ∪ {$C_{ffchild}$}})
   **return** $C_{G(ffcurr)}$

**cutset[]** *Merge* ($C_{G_1}$, $C_{G_2}$)
   **cutset** $C_{G_1 \cup G_2}[]$ = []
   **foreach** $c_i$ **in** $C_{G_1}$
      $N_{u,c_i}$ = *GetUpNodes*($c_i$,$G_1$)
      $N_{d,c_i}$ = *GetDownNodes*($c_i$,$G_1$)
      **foreach** $c_j$ **in** $C_{G_2}$
         $N_{u,c_j}$ = *GetUpNodes*($c_j$,$G_2$)
         $N_{d,c_j}$ = *GetDownNodes*($c_j$,$G_2$)
         **if** $N_{u,c_j} \cap N_{d,c_i}$ = Φ **and** $N_{d,c_j} \cap N_{u,c_i}$ = Φ **then**
            **cutset** $c_{union}$ = $c_i$ ∪ $c_j$
            $C_{G_1 \cup G_2}$ .*append*($c_{union}$)
   **return** $C_{G_1 \cup G_2}$

**Fig. 3.10**  Graph feedforward cutset identification pseudo code

found by recursively finding the set of feedforward cutsets for all the children of $ff_{root}$, which are then merged into a single set.

In Fig. 3.10, the *Merge* function merges different sets of feedforward cutsets and checks the validity of the union. In particular, let us consider two cutsets $c_i$ and $c_j$ of two sub-graphs spanning from two child nodes having a common parent as above. The union $c_{union}$ = $c_i$ ∪ $c_j$ is valid only if it is an actual cutset of the merged sub-graphs, i.e., if there is no common node between upstream and downstream nodes of $c_{union}$, by definition of cutset. In other words, $c_{union}$ is valid only if the intersection between the upstream nodes $N_{u,c_i}$ of $c_i$ with downstream nodes $N_{d,c_j}$ of $c_j$ is the empty set, and the same holds for the intersection of the downstream nodes $N_{d,c_i}$ of $c_i$ and the upstream nodes $N_{d,c_j}$ of $c_j$.

The above graph feedforward cutset identification method is exemplified in Fig. 3.11, where it is applied to the graph with non-linear pipeline in Fig. 3.8. In Fig. 3.11, the set of feedforward cutsets $C_{G(ff_3)}$ in the sub-graph spanning from $ff_3$ is obtained by merging feedforward cutsets in the sub-graphs spanning from the children nodes of $ff_3$ (i.e., nodes $ff_4$ and $ff_5$), which are $C_{G(ff_4)} \cup \{e_{34}\}$ (in blue) and

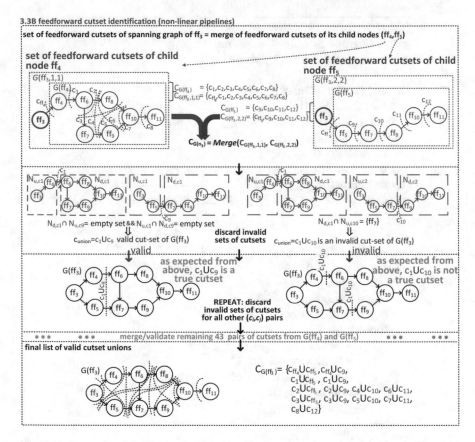

**Fig. 3.11** Example of feedforward cutset identification, based on the algorithm in Fig. 3.10 applied to the graph in Fig. 3.8

$C_{G(ff_5)} \cup \{e_{35}\}$ (in red). In turn, all the possible cutsets $C_{G(ff_4)}$ $\left(C_{G(ff_5)}\right)$ are found by recursively merging all possible feedforward cutsets of the sub-graphs spanning from the child nodes of $ff_4$ ($ff_5$) [33]. As first pair of merged cutsets from $C_{G(ff_4)}$ and $C_{G(ff_5)}$, the union of $c_1$ and $c_9$ is shown to be valid and is hence retained as possible feedforward cutset. Indeed, the intersection between the upstream nodes $N_{u,c_1}$ of $c_1$ with downstream nodes $N_{d,c_9}$ of $c_9$ is the empty set, and the same holds for the intersection of the downstream nodes $N_{d,c_1}$ of $c_1$ and the upstream nodes $N_{d,c_9}$ of $c_9$. Instead, the union of $c_1$ and $c_{10}$ is not a valid cutset as the intersection of $N_{d,c_1}$ and $N_{u,c_2}$ is $\{ff_7\}$ which is not an empty set. The above enumeration of possible cutsets of the graph spanned by $ff_3$ in Fig. 3.11 is repeated for the remaining 45 union cutsets. Among them, 13 are found to be valid cutsets and are listed on the bottom-right side of Fig. 3.11.

The complexity of finding a cutset through the above algorithm is $O(N \cdot N_{cuts} + M)$, where $N$ and $M$ are respectively the number of graph nodes and edges, whereas $N_{cuts}$ is the average number of cuts identified per node [33]. Hence, the resulting complexity is linear in the graph size. On the other hand, conventional brute-force (i.e., exhaustive) cutset search has exponential complexity $O(2^N)$, as the maximum number of possible sub-graphs of a graph with $N$ nodes is $2^N$. Executing the above algorithm on a server with Intel Xeon 5 processor and 128 GB RAM, cutset identification on a 4-bit FIR filter ($N = 100$, $M = 150$) and a 4-bit fixed-point multiplier ($N = 90$, $M = 180$) took about 1 s of CPU time, as opposed to more than 3,000 s under exhaustive search. For the more complex 16-bit FIR filter ($N = 2000$, $M = 3500$), the proposed cutset search algorithm required 30 s (in line with the expected linear complexity), whereas no result was generated under exhaustive search, as the system crashed down due to the memory over-usage.

## 3.6.3  Cutset-to-Pipeline Mapping (Step 3.3C)

The second step of Phase II in non-linear pipelines is to map the set of all possible feedforward cutsets from the previous subsection (Step 3.3B) onto actual register levels ranging from 1 (registers on the input side) to *pipedepth* (i.e., the maximum number of pipestages from any path going from inputs to any output). The feedforward cutsets identified in the previous step represent set of edges representing combinational logic between two successive flip-flops in the skeleton graph. The *cutset-to-pipeline mapping* Step 3.3C in Fig. 3.1a is needed to associate this set of edges (or cutsets) with the corresponding pipeline registers. Indeed, pipeline registers lying in such cutsets can be bypassed without modifying the overall functionality, as their bypass (or removal, in static microarchitectures) preserves the relative timing relationship of all signals at the input of any pipestage [31, 32].

   In general, flip-flops associated with the same register level always belong to the set of sources $S(c)$ of the same feedforward cutset $c$ [31, 32]. However, not all flip-flops belonging to $S(c)$ necessarily belong to the same register level. Indeed, by definition two registers belonging to the same register level cannot follow each other, i.e. there is no directed path that connects the two flip-flops in the graph. Hence, even if they both belong to $S(c)$, they are not in the same register level if one actually follows the other (see above and the example in Fig. 3.13 in red). Accordingly, flip-flops belonging to the same register level in non-linear pipelines are found by first considering the set $S(c)$, and then discard all nodes belonging to $S(c)$ having downstream nodes (i.e., the set of children $\{\cup children(n_i), n_i \epsilon S(c)\}$) overlapping with the set of upstream nodes $N_{u, c}$). In addition, $S(c)$ does not define a set of flip-flops in the same register level if $S(c)$ contains dummy nodes (i.e., not actual flip-flops). Accordingly, the nodes belonging to $S(c)$ of a given cutset $c$ are flip-flops belonging to the same register level if $\{\cup children(n_i), n_i \in S(c)\} \cap N_{u, c} \neq \emptyset$, and $S(c)$ does not contain dummy nodes. These considerations are summarized into the cutset-to-pipeline mapping algorithm in Fig. 3.12.

   The above cutset-to-pipeline mapping is exemplified in Fig. 3.13, as applied to the graph in Fig. 3.8. Analysis shows that there are 14 cutsets, although not all of them will actually define register levels, based on the above considerations. In particular, $S(c_6) = \{\text{ff}_4, \text{ff}_5\}$ is a set of flip-flops in the same register level since

**Fig. 3.12** Cutset-to-pipeline mapping pseudo code

## Algorithm: Cut-set to Pipeline

**pipeline** s[] = *Cutsets2Pipeline*($C_{G(ff_{root})}$)
**pipeline[]** *Cutsets2Pipeline* ($C_G$)
  **pipeline** s[] = []
  **for** i = 1: ($C_G$.*length*)
    **cutset** c = $C_G$[i]
    **if** {U child of $n_i$, $n_i \in S(c)$}$\cap N_{u,c}= \Phi$ && $S(c) \cap DN_G = \Phi$
      pipeline.*append*(S(c))
  **return** s

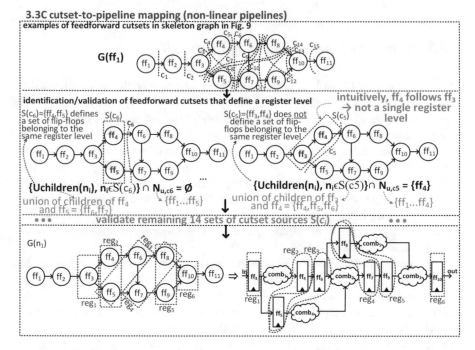

**Fig. 3.13** Example of cutset-to-pipeline mapping and resulting register levels $reg_1 \ldots reg_6$, based on the algorithm in Fig. 3.12 applied to the graph in Fig. 3.8

$\{\cup$children$(n_i), n_i \in S(c_6)\} \cap N_{u,c_6} = \varnothing$, and $S(c_6)$ does not contain any dummy node. In contrast, $S(c_5) = \{ff_3, ff_4\}$ does not define a set of flip-flops in the same register level, since $\{\cup$children$(n_i), n_i \in S(c_5)\} \cap N_{u,c_5} = \{ff_4\} \neq \varnothing$. Repeating the same arguments for the 14 cutsets in Fig. 3.13, only six of them turn out to define a set of flip-flops belonging to the same register level. In detail, $\{ff_1\}$ and $\{ff_2\}$ are immediately discarded because their sources contain dummy nodes. Mapping the nodes to the flip-flops of the original netlist in Fig. 3.8, the possible sets of flip-flops belonging to the same register level turn out to be $\{ff_3\}$, $\{ff_4, ff_5\}$, $\{ff_6, ff_5\}$, $\{ff_7, ff_8\}$, $\{ff_9, ff_8\}$, $\{ff_{10}\}$.

## 3.7 Automated Design Flow for Pipeline-Level Reconfiguration: Bypassable Registers Choice (Step 3, Phase III)

The outcome of Phase II in Sect. 3.6 for non-linear pipelines (Sect. 3.5 for linear pipelines) is a list of flip-flops organized into pipeline registers associated with progressive levels going from the inputs to the outputs. Such flip-flop organization into register levels allows to regain register transfer level understanding, which was previously lost in the original (or retimed) flattened gate-level netlist, as necessary to appropriately bypass registers.

Phase III of pipeline identification aims to identify the best possible choice of pipeline registers to be made bypassable (see flow in Fig. 3.1a). Such choice is determined by two main factors. First, the chosen registers should contain a number of flip-flops that is as large as possible. Indeed, this allows more substantial clocking energy savings in the shallow configuration, being the clock of the bypassable registers inhibited via clock gating. Second, bypassable registers should be inserted in pipestages whose merger in bypass mode leads to minimal clock cycle increase, compared to the re-pipelined (deep) microarchitecture without bypassable registers from Step II.

From the above considerations, the choice of bypassable registers is an optimization problem that aims to maximize the number of flip-flops to be bypassed, while keeping the clock cycle increase below a design target. Figure 3.14 summarizes the resulting algorithm to optimally choose the bypassable registers, among the multiple choices available from the registers identified in Phase II.

As first step discussed above, register levels having the largest number of flip-flops are shortlisted. In the example in Fig. 3.14 (step i), registers {$reg_4$, $reg_5$, $reg_6$, $reg_7$} are shortlisted, as they contain the largest number of flip-flops. As second step (step ii in Fig. 3.14), the above shortlisted register level having the lowest clock cycle increase is selected to be made bypassable. In the example in Fig. 3.14, the clock cycle increase is evaluated when $reg_4$, $reg_5$, $reg_6$, and $reg_7$ are respectively bypassed. The lowest maximum cycle of 1.6 ns is achieved when $reg_4$ is chosen as bypassable register, leading to the minimal cycle increase of 60% compared to the 1-ns cycle before bypassable register insertion. As third and last step (step iii in Fig. 3.14), the next iteration to add further bypassable register levels is prepared by eliminating the register levels that have been (partially or totally) absorbed in a previously chosen bypassable register level, as discussed above. In the example in Fig. 3.14, once $reg_4$ is chosen to be bypassed, $reg_5$ needs to be removed from the pool of the available register levels, since its node $ff_5$ has already been absorbed in $reg_4$.

The above procedure is repeated as long as there are additional register levels that can be made bypassable while meeting the targeted clock cycle, so that the clock energy in shallow configuration will be the lowest under the performance target. In the example of Fig. 3.14, the 1.6-ns critical path after the first above iteration is lower than the maximum targeted cycle of 2 ns. Hence, another iteration can be considered to add another bypassable register level. As in Fig. 3.14, the best choice is now the register level $reg_5$, leading to a 1.9-ns maximum cycle increase. Being the latter lower than the maximum targeted cycle, $reg_5$ is also made bypassable. The procedure then stops, as no further register level can be made bypassable without violating the 2-ns cycle target.

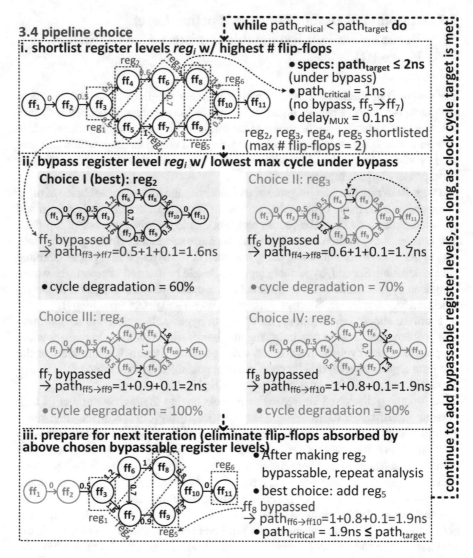

**Fig. 3.14** Example of optimal choice of registers to be made bypassable, based on the algorithm described in Sect. 3.6 applied to the graph in Fig. 3.8. The MUX timing penalty in bypassable flip-flops is assumed to be 0.1 ns in these examples

## 3.8   Automated Design Flow for Pipeline-Level Reconfiguration: Bypassable Register Replacement (Step 4)

Once the bypassable register levels are identified in Phase III (see Sect. 3.7), all flip-flops belonging to such levels are replaced by their bypassable version, as summarized in Step 4 in Fig. 3.1a. As shown in Fig. 3.15, bypassable register levels are replaced by simply manipulating the original gate-level netlist, overwriting the flip-

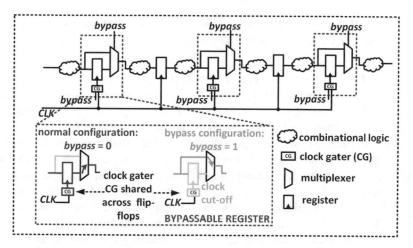

**Fig. 3.15** Selective replacement of flip-flops with bypassable flip-flops, and generation of final reconfigurable microarchitecture [26]

flop instance with the bypassable flip-flop in Fig. 3.15. The latter is added as a module comprising a flip-flop and a multiplexer with the same strength as the original flip-flop (which is easily created with structural description in Verilog, or other hardware description language).

As shown in Fig. 3.15, the *bypass* control signal is added as further input of the module under design, being controlled externally. In normal mode (*bypass* = 0), the flip-flops act as usual and leave all original pipestages intact, thus setting the microarchitecture in deep configuration. In bypass mode (*bypass* = 1), the multiplexer bypasses the flip-flop to make it transparent, merging the preceding and successive pipestages, thus leading to a shallower microarchitecture. To correctly optimize timing and meet the clock cycle specification in the (faster) deep configuration, the paths passing through the multiplexers are kept as false paths during synthesis and subsequent place and route. This also allows the synthesis and place&route tools to balance out pipestages, as the timing of the faster ones is invariably relaxed by CAD tool timing optimization, to reduce their energy contribution and area [29, 33]. As a side benefit, this naturally balances merged pipestages in the shallow microarchitecture as well.

As in Fig. 3.15, the clock signals of bypassable flip-flops are clustered and driven by clock gaters. To this aim, a single gater is introduced in the gate-level netlist to drive all clock signals of bypassable registers, before placement and routing. Then, the place&route tool automatically handles the buffer insertion to drive the clock pins of the bypassable flip-flops, according to the clock tree constraints. This suppresses clock activity and power both in bypassed flip-flops and in the related clock buffers, when the clock is inhibited in bypassable registers.

When the microarchitectural configuration needs to be changed, the pipeline is first flushed, then a new configuration is set, after which the pipeline restarts operation under the new configuration. The need for temporary interruption is not a significant limitation, considering the typically low number of cycles required by pipeline flushing, making the reconfiguration time negligible compared to the typical settling time of the supply voltage under dynamic voltage scaling [26].

## 3.9    Automated Design Flow Extension to Thread-Level Time-Interleaved Reconfiguration

The design flow for thread-level reconfiguration in processors is relatively easier in comparison to the pipeline-level one. As fundamental difference, the effectiveness of pipeline-level reconfiguration is limited in microarchitectures whose critical path lies in feedback paths, as the latter ones cannot be re-pipelined (see Sect. 3.3 and Fig. 3.4). Similarly, pipeline-level reconfiguration has limited applicability in architectures where re-pipelining introduces architectural hazards due to data inter-dependency (e.g., some computation depends on previous computations before it is completed, and stalls need to be inserted to resolve the hazards). Indeed, in this case the resolution of architectural hazards under re-pipelining requires a change in the control flow, and hence fundamental architectural (functional) changes that are unacceptable in applications with legacy requirements (e.g., binary-code software compatibility in microprocessors). In such cases or any other case where re-pipelining is difficult, pipeline-level reconfiguration is not an option, and thread-level must be introduced as discussed below.

In thread-level reconfiguration, logic depth is reduced by inserting additional registers to enable time-interleaved operation (see Chap. 2) and bypassing them to operate as in the original microarchitecture. Under time-interleaved operation, functionally equivalent tasks are simultaneously executed on multiple incoming input streams. This in turn enables the execution of multiple microprocessor threads (i.e., independent instruction steams and creation of multiple virtual cores), or processing of multiple input channels as in the case of multi-sensor platforms, among the many other examples.

The design flow for thread-level reconfiguration in Fig. 3.1b is structurally similar to the pipeline-level reconfiguration discussed in Sects. 3.3–3.8 and is somewhat simpler as discussed in the following. As shown in Fig. 3.16, the netlist generated at Step 1 in Fig. 3.1b is parsed to identify the flip-flops by simply tracing the clock port. This step also yields the *netlist graph* in the form of a hash table as described in Sect. 3.4.1. All the information associated with the gates and connection is stored as a graph database. Since this step operates directly on the gate-level netlist, the flow is architecture-agnostic and applies both to proprietary IP described at the RTL level or to existing IPs available in the form of a gate-level netlist (as usual for soft IPs provided by a third party, often times in an obfuscated gate-level form). In the following, the design flow is described assuming that up to two-thread operation is enabled, being its generalization to multiple threads immediate. At Step 1, each flip-flop is replaced by two cascaded flip-flops, as required by time interleaving in Chap. 2 (or, more in general, a number of cascaded flip-flops equal to the number of simultaneous threads).

The above-described Step 1 is analogous to Step 1 of the pipeline-level reconfiguration flow in Fig. 3.1a and is performed through a synthesis or P&R tool via a scripted set of commands. The replacement of each flip-flop by a pair of cascaded ones enables time interleaving of two completely independent input or instruction streams (i.e., two separate threads, as desired), as discussed in Chap. 2. At any given

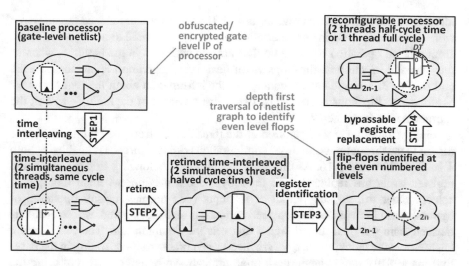

**Fig. 3.16** Pictorial description of the design flow in Fig. 3.1b to introduce thread-level reconfiguration via selective enablement of time interleaving in the gate-level netlist of a baseline design (e.g., processor). This diagram assumes dual-thread reconfiguration and can be readily adapted to any number of threads larger than two

cycle, odd-numbered stages process thread #1, whereas even-numbered stages process thread #2. This is fundamentally different from [26], which requires pipeline register insertion in appropriate points of the microarchitecture. At this step, no throughput improvement is achieved compared to the baseline design, since the clock frequency remains the same as the baseline. Indeed, the combinational logic delay in any odd pipestage has not changed, and the combinational logic delay in any even pipestage is zero.

To translate the register insertion into actual throughput improvement, the substantial delay imbalance between even and odd pipestages is evened out through retiming at Step 2 in Fig. 3.1b, or equivalently in Fig. 3.16. After retiming, the clock cycle is approximately halved compared to the baseline microarchitecture. The retimed netlist is now able to run at nearly doubled frequency, with thread #1 (thread #2) computations being completed in odd-numbered (even-numbered) cycles. This nearly preserves the same per-thread throughput, while doubling the thread count and hence the overall throughput. Such throughput increase would be unfeasible via simple re-pipelining, as the whole processor architecture would need to be completely revisited in terms of control flow (i.e., the original architecture would not be preserved, and very substantial architectural verification effort would be required).

Similar to the pipeline-level reconfiguration flow, retiming erases the original identity of the flip-flops and their associated register levels. Thus, registers to be bypassed need to identified so that the basic configuration will maintain exactly the functionality same as the baseline design. This step is analogous to the register identification Step 3 in Fig. 3.1a for pipeline-level reconfiguration, as shown in the thread-level flow in Figs. 3.1b and 3.16. In the thread-level reconfiguration flow, register identification is much simpler than pipeline-level since it simply needs to identify the registers at even register levels. The latter ones are indeed the ones that were added to the baseline microarchitecture, and hence need to be bypassed under single-thread operation to maintain the same operation as in the baseline, whereas they need to function as flip-flops when dual-thread operation is intended. At the end of Step 3, a new database containing the information about the register levels associated to flip-flops is created by depth first traversal of the *netlist graph* of the retimed netlist from Step 2.

Finally, dynamic reconfiguration of the thread count from the baseline with one thread to the throughput-enhanced configuration with two threads is enabled at Step 4 in Figs. 3.1b and 3.16. At this step, the flip-flops belonging to even-numbered register levels are replaced with their bypassable version, where a multiplexer selects either the register output or its input (i.e., normal register operation or bypass). The clock pin of the replaced registers is also clock gated, whose enable is derived from the dual-thread signal that sets the configuration. The clock gating ensures that the clocking energy is reduced when processor is operating in the baseline mode. This step is again performed through synthesis or P&R tool using the compiled set of scripts which is formed as an output of Step 3, with similar design considerations as in Sect. 3.8.

## 3.10   SRAM Reconfiguration at the Bank Level

### 3.10.1   Design Considerations on Memory Reconfiguration

In reconfigurable digital systems for extended power-performance range beyond voltage scaling, reconfiguration in memories is necessary as well, as discussed in the following. In the simplest instance of a microprocessor system with single-level memory hierarchy, memory is available in the form of a scratch pad or cache [29, 34–44]. Cache access for instruction fetch typically takes a few cycles (e.g., 1–4 from simple microcontrollers to high-performance processors), and any memory access time increase immediately translates into a performance degradation, either in terms of clock cycle increase or cycles per access. Also, the processor critical path often lies at the core-memory interface (e.g., instruction fetch) for non-trivial cache memory sizes, and data and instruction memory accesses are among the most energy-hungry operations in a microprocessor [35]. Hence, the core reconfiguration to improve either performance or energy efficiency easily makes the memory either the performance or the energy bottleneck, as discussed in the following.

In practical cases, three scenarios can occur in terms of relative speed of the baseline core and memory at nominal voltage: (a) the baseline memory is faster than the baseline core, (b) the baseline memory is slower than the baseline core, and (c) the baseline memory runs nearly at the same speed as the core. From an energy perspective, in the first scenario the voltage of the memory should be reduced to save dynamic and leakage energy at no throughput penalty. However, this scenario is rare as it typically occurs in systems with small memory size of a few KB, and high core logic depth of 100 $FO4$ or more [43]. Scenario (b) with the memory being slower than the core is much more common, and allows the core voltage to be down-scaled compared to the memory. Scenario (c) is most desirable in terms of both performance and energy efficiency (including leakage energy), as no part of cycle is unutilized by either the core or the memory. In scenarios (b) and (c), memory reconfiguration is needed to allow consistent memory-processor speed-up when the core is in a configuration offering higher performance than the baseline, as was discussed in previous subsections.

In the simple case of local, small-sized (e.g., kbit range) latch-based or flip-flop-based memories, their gate-level netlist can be easily re-pipelined or time-interleaved by reusing the same design flows discussed in the previous sections, to ultimately achieve consistent speed-up in both memory and logic. However, SRAM memories are more commonly adopted as local storage in view of their much better density. Unfortunately, pipeline-level reconfiguration is not an option in SRAMs, since their latency is tightly constrained by the core microarchitecture, and their physical design is tightly constrained by density considerations. Also, under both pipeline- and thread-level reconfiguration of the microprocessor core, the SRAMs need to have nearly halved access time to sustain full-speed core operation, compared to a baseline design with no re-pipelining or a single thread. Hence, a different reconfiguration approach and flow need to be introduced in SRAMs to selectively improve their access time beyond nominal voltage, whenever the core configuration requires it.

In practical designs, SRAMs are mostly designed through memory compilers to mitigate their otherwise substantial design effort. Hence, SRAM configuration needs to nicely fit an existing compiled memory and requires minimal additional design effort. Accordingly, in the following design, techniques to modify compiled SRAM macros are discussed for the array and the periphery. In particular, the concept of row aggregation for speed-up (see Chap. 2) is applied in a selective manner, enabling it when shorter access time is needed.

### 3.10.2   Background on Low-Power and Reconfigurable Memories

Small-signal (i.e., low-swing) bitline sensing is ubiquitously adopted in SRAMs to achieve fast read access. Having the sensed bitline voltage a low swing, the offset of the sense amplifier has a detrimental effect in terms of read robustness at a given swing or requires larger swing (i.e., slower access) for a given robustness. Accordingly, several techniques to mitigate the sense amplifier offset have been introduced. For example, in [45] an on-chip offset monitor is introduced to accurately characterize the sense amplifier offset at testing time, and hence avoid overdesign (see Fig. 3.17a). In [46], the offset was reduced by using two smaller sense amplifiers, so that the replica with smaller offset can be selected at testing time (see Fig. 3.17b). This is shown to increase the probability to achieve a smaller offset, compared to the adoption of a single larger sense amplifier with lower offset standard deviation. In [47], sense amplifier replicas are instead combined by allowing input swapping, so that the replica offsets can be either added or subtracted (see Fig. 3.17c). This offers more opportunities to reduce the offset compared to the selection of one replica, achieving 3X sense amplifier offset reduction. In [48], the sense amplifier is reconfigured to pre-amplify the bit-line differential voltage (see Fig. 3.17d) and is then reconfigured back as to a latch, achieving 42% read speed improvement.

A wide range of other techniques to improve the SRAM read/write speed at given robustness (or enhance the latter at given speed) have been extensively used in commercial SRAMs, such as word line boosting [46], reducing the number of cells per bitline, hierarchical bitline [49], charge-sharing techniques for voltage swing reduction in bitlines [50]. Such tradeoffs have been made more favorable by leveraging the error resilience of several prominent applications (e.g., machine learning, computer vision, physical signal processing) and introducing the concept of energy-quality scalable SRAMs [51]. In energy-quality scalable SRAMs, additional energy on MSBs is purposely paid for to improve their resiliency against aggressive voltage scaling (e.g., negative bitline boosting, non-uniform Error Correcting Codes favoring MSBs over LSBs), while not spending any additional energy on LSBs. These techniques can assure errorless operation when applied uniformly to all bits and allow graceful quality degradation when selectively applied to different bit positions [51]. The resulting quality degradation mitigation allows more aggressive voltage scaling and energy savings that are superior compared to simple approximate SRAMs [52].

When the supply voltage is reduced, the access time degradation in SRAMs is well known to be worse than the one in the processor core. This is due to both the stronger sensitivity of the bitcell on-current to the voltage compared to logic gates and the larger timing margin increase due to process variations (since there is minimal or no averaging effect across the few circuits along the SRAM critical path, as opposed to the critical logic path with typically tens of gates). At the same time, the minimum operating voltage $V_{min}$ of SRAMs is generally in the above-threshold region (e.g., at least 0.6–0.7 V), which is significantly higher than the $V_{min}$ of logic that lies in the near- or sub-threshold region (often below 0.4 V). The different performance sensitivity to the supply voltage and the different $V_{min}$ impose that the SRAM and the processor core are in different voltage domains. This indeed mitigates or eliminates the memory-processor performance mismatch, when operating above $V_{min}$. Under aggressive voltage scaling, the SRAM cannot operate below its $V_{min}$, whereas the core allows further voltage scaling. Hence, at the lower end of the voltage range and performance, the SRAM becomes faster than the processor core, as opposed to the higher end of the voltage range.

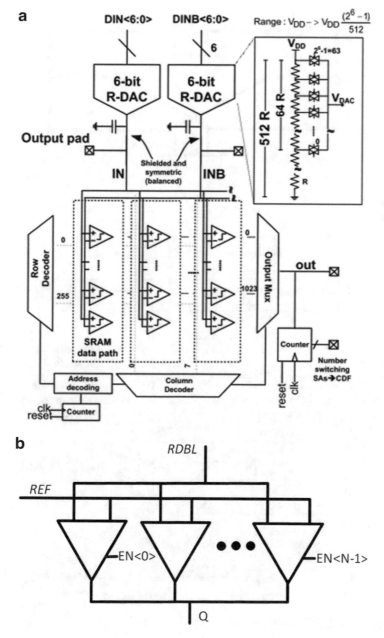

**Fig. 3.17** Examples of sense amplifier designs with improved offset: (**a**) sense amplifier offset monitor [45], (**b**) redundancy in sense amplifier design [46], (**c**) reconfigurable sense amplifier [47], (**d**) reconfigurable SRAM with bitline pre-amplification [48]

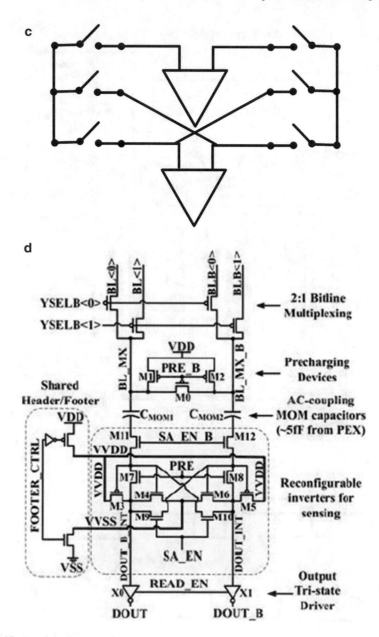

**Fig. 3.17** (continued)

### 3.10.3   Row Aggregation Technique and Reconfiguration for Selective Performance Enhancement Beyond Nominal Voltage

As mentioned in Chap. 2, the performance of SRAM memories can be improved beyond allowed at nominal voltage by reconfiguring its architecture. This is accomplished by introducing a dual-wordline architecture to flexibly increase the effective on-current of the bitcell, when performance needs to be boosted. In this technique, the bitcell read current $I_{read}$ is doubled by aggregating two bitcells to read the same data, activating their wordlines at the same time instead of selecting only one. Similar considerations hold on the data written into the two bitcells, as their content needs to be the same during the next read, and hence requires a previous simultaneous write of the same data on both bitcells. When instead normal operation is targeted (i.e., no architectural speed-up is needed), only one wordline is conventionally activated at a time. Figure 3.18 shows the architecture of row-aggregated array. When the dual-wordline activation signal $ROWAGG$ is asserted (deasserted), each bitline is pulled down by two bitcells (one bitcell). Thus, the average read current $I_{read}$ of the bitcell increases by a factor of two (does not increase), speeding up the bitline voltage development by the same factor compared to a baseline SRAM design with single wordline activation.

The only required form of reconfiguration to operate in the two above modes needs to be inserted into the row decoder, as it needs to activate one wordline at a time when $ROWAGG = 0$ or a pair when $ROWAGG = 1$. When $ROWAGG = 1$ (i.e., row aggregation mode), the data stored in the simultaneously accessed cells has to be the same, thus halving the memory capacity. In other words, the doubled speed of the bitline voltage development is achieved at the expense of doubled area or halved memory capacity. Overall, the resulting reconfigurable memory has an elastic capacity, i.e., full capacity in normal mode, half capacity in row aggregation mode.

### 3.10.4   Embedding Reconfigurable Row Aggregation Through Minor Modifications of Existing Compiled Memories

As shown in Fig. 3.20, two decoder wordlines need to be simultaneously activated when the array operates in row aggregation mode, otherwise only one in normal mode. This feature can be easily introduced in existing row decoders Row decoder at insignificant area overhead, as discussed below.

Equation (3.1) shows the Boolean expression describing the operation of a conventional 8-to-256 (0...255) row decoder, and in particular the wordline activation signal for row #128 (it needs to be asserted when the input address is $A_7...A_0 = (128)_{10} = 10000000$):

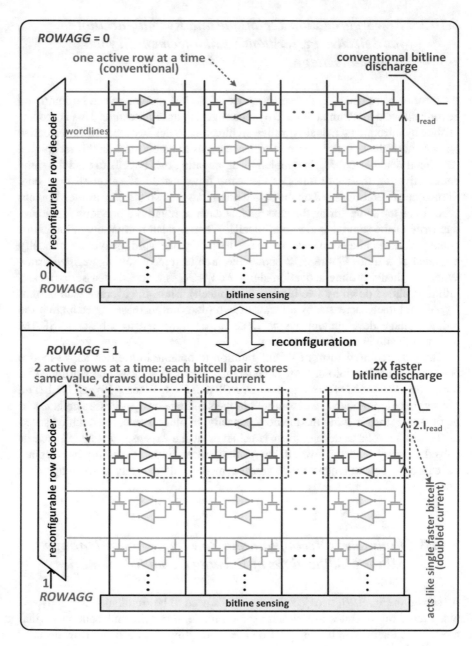

**Fig. 3.18** SRAM row aggregation doubles the bitline discharge speed thanks to the simultaneous activation of two wordlines (with each pairs of corresponding bitcells storing the same value), while leaving it unaltered when a single wordline is activated

$$WL_{128} = A_7.\overline{A_6}.\overline{A_5}.\overline{A_4}.\overline{A_3}.\overline{A_2}.\overline{A_1}.\overline{A_0} \tag{3.1}$$

The above equation describes the activation of $WL_{128}$ in a conventional decoder, and hence the reconfigurable decoder in normal mode. Instead, in row aggregation mode $WL_{128}$ needs to be simultaneously asserted with its paired row. In the following, it is assumed by convention that each row is unambiguously paired with the one differing only for its most significant bit (MSB) $A_7$, which is complemented (i.e., obtained by subtracting 128 from the row it is paired with). In other words, in row aggregation mode, the memory sub-bank address space is divided into two halves as in Fig. 3.19, which are separately addressable in normal mode ($ROWAGG = 0$) and are instead simultaneously addressed by activating the pair of rows having the same position within each of half sub-bank in row aggregation mode ($ROWAGG = 1$). Accordingly, the Boolean expression of $WL_{128}$ in the reconfigurable memory decoder is

$$WL_{128} = A_7 \cdot \overline{A_6}.....\overline{A_0}.\overline{ROWAGG} + \overline{A_7} \cdot \overline{A_6}.....\overline{A_0}.ROWAGG \tag{3.2}$$

which can be rewritten as

$$WL_{128} = \left( A_7 \oplus ROWAGG \right) \cdot \overline{A_6}.....\overline{A_0} \tag{3.3}$$

where it was observed that $A_7 \cdot \overline{ROWAGG} + \overline{A_7} \cdot ROWAGG$ is the XOR of $A_7$ and $ROWAGG$.

The above consideration and derivations can be immediately extended to any address within the same sub-bank in Fig. 3.19. Accordingly, the reconfigurable decoder is obtained from a conventional SRAM sub-bank decoder by simply replacing its MSB $A_7$ by ($A_7 \oplus ROWAGG$) as in Eq. (3.3). From an implementation viewpoint, this replacement simply requires the substitution of the conventional inverter driving $A_7$ in the pre-decoder stage by an XOR gate. This is shown in Fig. 3.20,

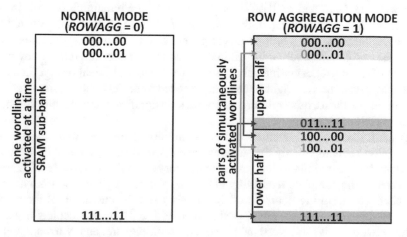

**Fig. 3.19** Reconfigurable SRAM architecture: organization of the memory sub-bank address space in normal and row aggregation mode

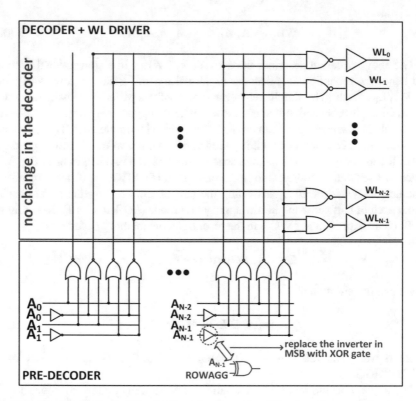

**Fig. 3.20** Reconfigurable row decoder Row decoder and modification of a conventional decoder to embed selective row aggregation

where the general decoder gate-level implementation is shown for a generic sub-bank with an $N$-bit address (instead of $N = 8$ adopted in the above example). The inverter in the pre-decoder in red color is replaced by an XOR gate combining its address MSB $A_{N-1}$ and *ROWAGG* as in Eq. (3.3). The remaining part of decoder remains completely unchanged, thus allowing full reuse of the pitch-matched part of the decoder and the wordline driver generated by the memory compiler. This keeps the design effort very minimal in the decoder and does not require any modification of the most design-intensive part of the decoder physical design (i.e., pitch matching, given the very tight bitcell pitch needed to achieve the required density). In terms of overhead, this replacement comes at insignificant area penalty compared to the overall array area.

Overall, the array arrangement of the reconfigurable memory with selective row aggregation remains exactly the same as the SRAM macro generated by the memory compiler. The minimal changes in the memory to implement row-level aggregation ensures the complete reusability of all the peripherals generated from memory compiler. The resulting floorplan of the memory and the allocation of the address space to the memory sub-banks is shown in Fig. 3.21, which depicts a generic memory array with $L$ bit word, $M : 1$ column multiplexing, and $N$ rows automatically generated using memory compiler.

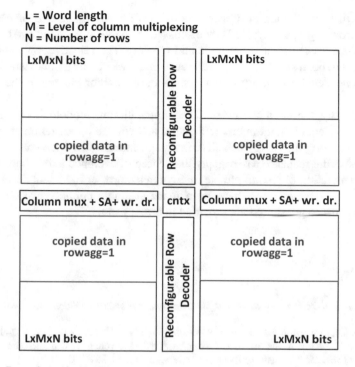

**Fig. 3.21**  Reconfigurable SRAM array with row aggregation and its arrangement into sub-banks

## 3.11   Conclusion

In this chapter, automated design methodologies have been introduced to extend the power-performance tradeoff beyond voltage scaling and to enhance energy efficiency across a wide voltage range. Such methodologies have been developed for both logic with pipeline- and thread-level reconfiguration, as well as for SRAM arrays. The common goal has been to introduce drop-in solutions for existing architectures that allow the above capability at very low design effort.

Regarding the design methodologies for logic, they are based on gate-level manipulations and are hence architecture-agnostic and can be adopted regardless of whether the design is described at the RTL level or it is externally provided by a soft IP vendor. Such methodologies rely on commercial EDA tools and scripts that implement graph algorithms to integrate tools into automated design flows. In particular, a graph algorithm to cluster flip-flops into ordered pipeline stages in a flattened gate-level netlist has been introduced, along with a timing-aware strategy to identify which registers need to be made bypassable. Pipeline-level reconfiguration has been shown to be well suited for application-specific hardware and accelerators, whereas thread-level is generally applicable to general-purpose architectures such as microprocessors. Regarding SRAMs, low-effort design methodologies to selectively boost up the performance beyond allowed by nominal voltage have been introduced. In particular, design methodologies for reconfigurable SRAMs with selective row aggregation have been discussed to enable synergistic and consistent performance boost of both core logic and memory. The selective row aggregation technique has been shown to allow full reuse of the vast majority of existing SRAM arrays generated by memory compilers, as only a minor change in the decoder is needed.

In summary, the microarchitectural reconfiguration methodologies in this chapter ultimately enable joint microarchitecture and voltage optimization at run time, further improving the energy benefits of voltage scaling under wide voltage scaling.

To enable the reader to easily apply the concepts developed in this book towards further exploration of reconfigurable microarchitectures or EDA tool development, all the above design scripts have been made available publicly in [53]. Further details on how to get started are provided in the Appendix.

# References

1. *International Technology Roadmap for Semiconductors: 2015 edition.* http://www.itrs.net, (2013)
2. B. Nikolić, Power-limited design, in *Proceedings of ICECS 2007*, (2007), pp. 927–930
3. T. Burd, T. Pering, A. Stratakos, R. Brodersen, A dynamic voltage scaled microprocessor system, in *IEEE ISSCC Digest of Technical Papers*, (2015), pp. 294–295
4. S. Jain et al., A 280mV-to-1.2V wide-operating-range IA-32 processor in 32 nm CMOS, in *IEEE ISSCC Digest of Technical Papers*, (2012), pp. 66–67
5. W. Wang, P. Mishra, System-wide leakage-aware energy minimization using dynamic voltage scaling and cache reconfiguration in multitasking systems. IEEE Trans. VLSI Syst. **20**(5), 902–910 (2012)
6. A. Chandrakasan, D. Daly, D. Finchelstein, J. Kwong, Y. Ramadass, M. Sinangil, V. Sze, N. Verma, Technologies for ultradynamic voltage scaling. Proc. IEEE **98**(2), 191–214 (2010)
7. M. Seok, D. Jeon, C. Chakrabarti, D. Blaauw, D. Sylvester, Extending energy-saving voltage scaling in ultra low voltage integrated circuit designs, in *Proc. of ICICDT 2012—IEEE International Conference on Integrated Circuit Design and Technology*, (2012), pp. 2–5
8. D. Jacquet, F. Hasbani, P. Flatresse, R. Wilson, F. Arnaud, G. Cesana, P. Magarshack, A 3 GHz dual core processor ARM cortex TM -A9 in 28 nm UTBB FD-SOI CMOS with ultra-wide voltage range and energy efficiency optimization. IEEE J. Solid State Circuits **49**(4), 812–826 (2014)
9. F. Abouzeid, S. Clerc, B. Pelloux-Prayer, F. Argoud, P. Roche, 28nm CMOS, energy efficient and variability tolerant, 350 mV-to-1.0 V, 10 MHz/700 MHz, 252 bits frame error-decoder, in *Proceedings of ESSCIRC 2012, Bordeaux, France*, (2012), pp. 153–156
10. S. Hsu, A. Agarwal, M. Anders, S. Mathew, H. Kaul, F. Sheikh, R. Krishnamurthy, A 280 mV-to-1.1 V 256b reconfigurable SIMD vector permutation engine with 2-dimensional shuffle in 22 nm CMOS, in *ISSCC Digest of Technical Papers, San Francisco (CA)*, (2012)
11. S. Hanson, B. Zhai, K. Bernstein, D. Blaauw, A. Bryant, L. Chang, K.K. Das, W. Haensch, E.J. Nowak, D.M. Sylvester, Ultralow-voltage, minimum-energy CMOS. IBM J. Res. Dev. **50**(4/5) (2006)
12. S. Hanson, B. Zhai, D. Blaauw, D. Sylvester, A. Bryant, X. Wang, Energy optimality and variability in subthreshold design, in *Proceedings of ISLPED 2006*, pp. 363–365

13. W. Zhao, Y. Ha, M. Alioto, Novel self-body-biasing and statistical design for near-threshold circuits with ultra energy-efficient AES as case study. IEEE Trans. VLSI Syst. **23**(8), 1390–1401 (2015)
14. Y. Zhang, M. Khayatzadeh, K. Yang, M. Saligane, M. Alioto, D. Blaauw, D. Sylvester, iRazor: 3-transistor current-based error detection and correction in an ARM Cortex-R4 Processor, in *IEEE ISSCC Digest of Technical Papers*, (2016), pp. 160–161
15. K. Nose, T. Sakurai, Optimization of $V_{DD}$ and $V_{TH}$ for low-power and high-speed applications, in *Proceedings of DAC, Yokohama (Japan)*, (2000)
16. B. Zhai, D. Blaauw, D. Sylvester, K. Flautner, Theoretical and practical limits of dynamic voltage scaling, in *Proceedings of DAC*, (2004)
17. S. Jain, L. Lin, M. Alioto, Design-oriented energy models for wide voltage scaling down to the minimum energy point. IEEE Trans. CAS Pt. I **64**(12), 3115–3125 (2017)
18. A.P. Chandrakasan, S. Sheng, R.W. Brodersen, Low-power CMOS digital design. IEEE J. Solid State Circuits **27**(4), 473–484 (1992)
19. M. Alioto, Ultra-low power VLSI circuit design demystified and explained: a tutorial. IEEE Trans. Circuits Syst. I Regul. Pap. **59**(1), 3–29 (2012)
20. H. Shimada, H. Ando, T. Shimada, Pipeline stage unification: a low-energy consumption technique for future mobile processors. Proc. Int. Sympos. Low Power Electr. Design **2003**, 326–329 (2003)
21. A. Efthymiou, J.D. Garside, Adaptive pipeline depth control for processor power-management, in *Proceedings of IEEE International Conference on Computer Design: VLSI in Computers and Processors*, (2002), pp. 454–457
22. S. Vijayalakshmi, A. Anpalagan, I. Woungang, D.P. Kothari, Power management in multi-core processors using automatic dynamic pipeline stage unification, in *2013 International Symposium on Performance Evaluation of Computer and Telecommunication Systems (SPECTS), Toronto (Canada)*, (2013), pp. 120–127
23. S. Chellappa, C. Ramamurthy, V. Vashishtha, L.T. Clark, Advanced encryption system with dynamic pipeline reconfiguration for minimum energy operation, in *Proceedings of 16th International Symposium on Quality Electronic Design (ISQED), Santa Clara (CA)*, (2015), pp. 201–206
24. H. Jacobson, Improved clock-gating through transparent pipelining, in *Proceedings of the International Symposium on Low Power Electronics and Design 2004, Newport Beach (CA)*, (2004), pp. 26–31
25. S. Manne, A. Klauser, D. Grunwald, Pipeline gating: speculation control for energy reduction, in *Proceedings of 25th Annual International Symposium on Computer Architecture, Barcelona (Spain)*, (1998), pp. 132–141
26. S. Jain, L. Lin, M. Alioto, Dynamically adaptable pipeline for energy-efficient microarchitectures under wide voltage scaling. IEEE Journal of Solid-State Circuits **53**(2), 632–641 (2018)
27. S. Jain, L. Lin, M. Alioto, Automated design of reconfigurable microarchitectures for accelerators under wide voltage scaling. In print on *IEEE Transactions on Very Large Scale Integration Systems*
28. S. Jain, L. Lin, M. Alioto, Drop-in energy-performance range extension in microcontrollers beyond VDD scaling, in *2019 IEEE Asian Solid-State Circuits Conference, Macau*, (2019), pp. 125–128
29. Synopsys, *Design Compiler User Manual Version X-2005*. Accessed 9 September 2005
30. Cadence, *Encounter™ User Guide Product Version 4.1.5*. Accessed May 2005
31. D. Markovic, R.W. Brodersen, *DSP Architecture Design Essentials* (Springer, Berlin, 2012)
32. K. Parhi, *VLSI Digital Signal Processing Systems: Design and Implementation* (Wiley, New York, 1999)
33. S. Chatterjee, On algorithms for technology mapping, in *Technical Report No. UCB/EECS-2007-100*, http://www.eecs.berkeley.edu/Pubs/TechRpts/2007/EECS-2007-100.html, Accessed 16 August 2007

34. J.L. Hennessy, D.A. Patterson, *Computer Architecture: A Quantitative Approach*, 6th edn. (Morgan Kaufmann, San Francisco, CA, 2019)
35. M. Gautschi et al., Near-threshold RISC-V core with DSP extensions for scalable IoT endpoint devices. IEEE Trans. Very Large Scale Integr. Syst. **25**(10), 2700–2713 (2017)
36. M. Alioto (ed.), *Enabling the Internet of Things—From Integrated Circuits to Integrated Systems* (Springer, Berlin, 2017)
37. M. Alioto, E. Consoli, G. Palumbo, *Flip-Flop Design in Nanometer CMOS—From High Speed to Low Energy* (Springer, Berlin, 2015)
38. D. Chinnery, K. Keutzer, *Closing the Power Gap Between ASIC & Custom* (Springer, Berlin, 2007)
39. V. Srinivasan et al., Optimizing pipelines for power and performance, in *Proceedings of International Symposium on Microarchitectures*, (2002), pp. 333–344
40. V. Zyuban, D. Brooks, V. Srinivasan, M. Gschwind, P. Bose, P.N. Strenski, P.G. Emma, Integrated analysis of power and performance for pipelined microprocessors. IEEE Trans. Comput. **53**(8), 1004–1016 (2004)
41. N. Weste, D. Harris, *CMOS VLSI Design: A Circuits and Systems Perspective*, 4th edn. (Addison-Wesley, New York, 2011)
42. H. Shimada, H. Ando, T. Shimada, A hybrid power reduction scheme using pipeline stage unification and dynamic voltage scaling, in *Proceedings of IEEE COOL Chips*, (2006), pp. 201–214
43. J. Myers, A. Savanth, R. Gaddh, D. Howard, P. Prabhat, D. Flynn, A subthreshold ARM cortex-M0+ subsystem in 65 nm CMOS for WSN applications with 14 Power Domains, 10T SRAM, and integrated voltage regulator. IEEE J. Solid State Circuits **51**(1), 31–44 (2016)
44. Y. Zhang, L. Xu, Q. Dong, J. Wang, D. Blaauw, D. Sylvester, Recryptor: a reconfigurable cryptographic cortex-M0 processor with in-memory and near-memory computing for IoT security. IEEE J. Solid State Circuits **53**(4), 995–1005 (2018)
45. M.H. Abu-Rahma et al., Characterization of SRAM sense amplifier input offset for yield prediction in 28 nm CMOS, in *Proceedings of the Custom Integrated Circuits Conference*, (2011)
46. N. Verma, a.P. Chandrakasan, A 256 kb 65 nm 8T subthreshold SRAM employing sense-amplifier redundancy. IEEE J. Solid State Circuits **43**(1), 141–149 (2008)
47. M. Khayatzadeh, F. Frustaci, D. Blaauw, D. Sylvester, M. Alioto, A reconfigurable sense amplifier with 3X offset reduction in 28nm FDSOI CMOS, in *IEEE Symposium on VLSI Circuits, Digest of Technical Papers*, vol. 2015, (2015), pp. C270–C271
48. B. Giridhar, N. Pinckney, D. Sylvester, D. Blaauw, A reconfigurable sense amplifier with auto-zero calibration and pre-amplification in 28nm CMOS. IEEE Int. Solid State Circuits Conf. Dig. Tech. Pap. **57**, 242–243 (2014)
49. M. Yoshimoto et al., A divided word-line structure in the static RAM and its application to a 64K full CMOS RAM. IEEE J. Solid State Circuits **18**(5), 479–485 (1983)
50. T.W. Oh, H. Jeong, J. Park, S.O. Jung, Pre-charged local bit-line sharing SRAM architecture for near-threshold operation. IEEE Trans. Circuits Syst. I Regul. Pap. **64**(10), 2737–2747 (2017)
51. F. Frustaci, M. Khayatzadeh, D. Blaauw, D. Sylvester, M. Alioto, SRAM for error-tolerant applications with dynamic energy-quality management in 28 nm CMOS. IEEE J. Solid State Circuits **50**(5), 1310–1323 (2015)
52. M. Alioto, V. De, A. Marongiu, Energy-quality scalable integrated circuits and systems: continuing energy scaling in the Twilight of Moore's Law. IEEE J. Emerg. Select. Topics Circuits Syst. **8**(4), 653–678 (2018)
53. M. Alioto, S. Jain, *RECMICRO: Design Framework and Scripts to Design Reconfigurable Microarchitectures* [Online], http://www.green-ic.org/recmicro

# Chapter 4
# Case Studies of Reconfigurable Microarchitectures: Accelerators, Microprocessors, and Memories

**Abstract** In this chapter, several case studies of reconfigurable microarchitectures are presented and enriched with testchip measurement and simulation results. The test vehicles were designed according to the principles in Chap. 2 and the automated methodologies developed in Chap. 3. Case studies cover pipestage- and thread level reconfiguration for accelerators and microprocessors, as well as bank-level for SRAMs. Typical energy efficiency and power-performance range gains over conventional voltage scaling are discussed, along with opportunities for further improvements. In particular, case studies include arithmetic building blocks (e.g., multipliers), digital filters and FFT accelerators, commercial microcontrollers, and SRAMs. To appreciate the effect of the underlying design decisions, various versions of these designs are either demonstrated on silicon or in simulation and compared.

**Keywords** Reconfigurable microarchitecture · Reconfigurable SRAM · Dynamic energy · Leakage energy · Above-threshold region · Near-threshold region · Sub-threshold region · Fixed microarchitectures · Minimum energy point (MEP) · Dynamically adaptable pipelines · Dynamic voltage frequency scaling · Bypassable register · Retiming · Fast Fourier Transform (FFT) · Fixed-point multiplier · Finite impulse response (FIR) filter · Microcontroller · Microcontroller unit (MCU) · ARM Cortex-M0 · Data memory · Program memory · Testchip · Post-layout simulations · Timing overhead · Area overhead · Energy overhead · Beyond-voltage scaling throughput enhancement · Beyond-voltage scaling energy reduction · Time interleaving · SRAM · Row decoder · FIFO · Twiddle factor · Look-up table (LUT) · Computational element · Commutator · Dragonfly · Complex multiplier · Logic depth · Bitline · Wordline · Memory sub-bank · Deep configuration · Shallow configuration · Process variations · Variability · Voltage margin · Statistical characterization of process variations · Access time · Row aggregation · Clock tree · Sequential cells · Latch · Energy sensitivity · Translator logic · AHB bus · XOR gate · $V_{min}$ · Minimum voltage

© Springer Nature Switzerland AG 2020
S. Jain et al., *Adaptive Digital Circuits for Power-Performance Range beyond Wide Voltage Scaling*, https://doi.org/10.1007/978-3-030-38796-9_4

In this chapter, several case studies of reconfigurable microarchitectures are presented and enriched with testchip measurement and simulation results. The test vehicles were designed according to the principles in Chap. 2 and the automated methodologies developed in Chap. 3. Case studies cover pipestage- and thread-level reconfiguration for accelerators and microprocessors, as well as bank-level for SRAMs. Typical energy efficiency and power-performance range gains over conventional voltage scaling are discussed, along with opportunities for further improvements. In particular, case studies include arithmetic building blocks (e.g., multipliers), digital filters and FFT accelerators, commercial microcontrollers, and SRAMs. To appreciate the effect of the underlying design decisions, various versions of these designs are either demonstrated on silicon or in simulation and compared.

## 4.1   Fast Fourier Transform (FFT) Accelerator

### 4.1.1   Microarchitecture and Design of Its Dynamically Adaptable Pipeline Counterpart

The concept of dynamically adaptable pipelines was applied to radix-4, 256-point, 16-bit input, fixed-point complex FFT engine with a modified MDC architecture [1] in 40 nm. The testchip micrograph is shown in Fig. 4.1a, and its main microarchitectural features are summarized in Fig. 4.1b.

The adopted architecture of the FFT accelerator is depicted in Fig. 4.2 and was implemented with both a conventional static microarchitecture with fixed pipeline depth and a dynamically adaptable pipeline. The design flow in Fig. 3.1a was employed to automatically generate the dynamically adaptable pipeline from the gate-level netlist obtained from the conventional design.

The throughput target was set to test the capability of dynamically adaptable pipelines to deliver high throughputs exceeding the highest compared to prior art, while still achieving the lowest energy when reconfigured. Accordingly, the subranges achieved by state-of-the-art FFT accelerators were concatenated into a single one [1–4] and were further extended from ~1 MS/s to ~8 GS/s. The adopted FFT architecture delivers eight FFT transforms per cycle, hence the maximum throughput demands a clock period of approximately 1 ns, or equivalently a logic depth of 33FO4/pipestage at the nominal voltage of 1.1 V. Such maximum throughput was targeted at the nominal voltage since achieving the maximum throughput at lower voltages would inevitably increase both clocking and combinational energy, due to the tighter timing constraints. In turn this would limit the energy gains of microarchitecture reconfiguration, in addition to its limitations at logic depths approaching the 20FO4/pipestage range (see detailed discussion on logic depth in Sect. 2.1.1).

The critical path of the architecture in Fig. 4.2 lies in the basic FFT computational element CELUT, which contains a fixed-point complex multiplier, three fixed-point complex adders (CE), and a look-up table (LUT). Accordingly, bypassable registers were inserted in the CELUT block. As expected, the FFT architecture is highly regu-

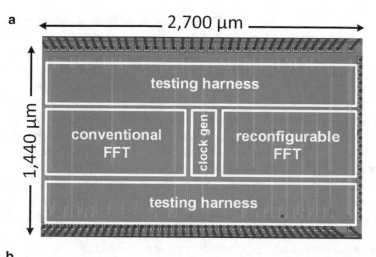

**a** ← 2,700 µm →

1,440 µm

testing harness

conventional FFT

clock gen

reconfigurable FFT

testing harness

**b**

| | conventional FFT | reconfigurable FFT | |
|---|---|---|---|
| gate count | 182k | 230k | |
| area | 0.53mm² | 0.53mm² | |
| technology | 40nm LP | 40nm LP | |
| $V_{min}$ | 0.34V | 0.34V | |
| max. frequency @ 1.1V | | deep | shallow |
| | 657MHz | 971 MHz | 571 MHz |
| logic depth | 60 *FO4* | 33 *FO4* | 64 *FO4* |

**Fig. 4.1** (**a**) 40 nm FFT testchip micrograph, (**b**) essential data on its microarchitectural features

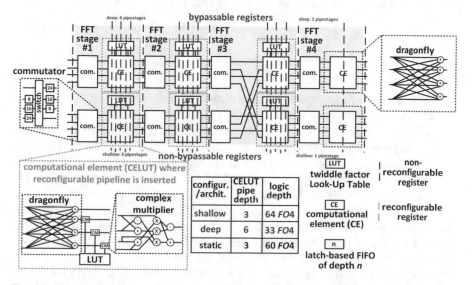

| configur. /archit. | CELUT pipe depth | logic depth |
|---|---|---|
| shallow | 3 | 64 *FO4* |
| deep | 6 | 33 *FO4* |
| static | 3 | 60 *FO4* |

**Fig. 4.2** Adopted FFT architecture and related features in the static and dynamically adaptable pipeline. The vertical dashed and thick lines represent the registers (bypassable ones are in red) and describe the arrangement into pipestages in both deep and shallow configurations

lar and includes a CELUT in every FFT stage from Fig. 4.2, excepting the last one. The latter does not include the LUT and the complex multiplier and is hence simpler and faster than all other stages by a factor that turned out to be approximately 3×.

A first step of the design flow described in Chap. 3 (Step 1 in Fig. 3.1a), the original CELUT was re-pipelined to achieve a deep fixed microarchitecture pipeline with 33FO4/pipestage, to meet the above discussed maximum throughput target. As a result, the CELUT in the first, second, and third FFT stage were re-pipelined with six pipestages. To achieve the same target, the last one was arranged with only two pipeline stages as shown in Fig. 4.2, in view of its 3× lower logic depth. After defining the initial fixed deep pipeline (Step 1 in Fig. 3.1a) and retiming it (Step 2), selected conventional registers were turned into bypassable registers according to Steps 3–4. After bypassing registers, the logic depth in the shallow configuration turned out to be 64FO4 per stage, which is about twice the logic depth of the deep configuration as expected. The ratio of the pipestage delay of the deep and shallow microarchitecture is not exactly two because of the slightly different gate sizing. In turn, this is due to the presence of the additional multiplexers in bypassable registers, and the discrete nature of the optimization performed by the synthesis tool.

To create a baseline design to evaluate the benefits and penalties of dynamically adaptable pipelines, a conventional FFT engine with static microarchitecture was designed to achieve approximately the same logic depth as the shallow configuration (i.e., 64FO4). The effective logic depth of the fixed pipeline design results to 60FO4 and is 4FO4 lower than the shallow configuration of the dynamically adaptable pipelined design. Again, this is due to the presence of multiplexers and the discrete synthesis optimization, as discussed above.

The test vehicle in shallow and deep configuration was verified against the original shallow and deep fixed microarchitecture through dynamic verification (i.e., checking the correct match of the output test vectors).

### 4.1.2  Measurement Results on a Single Die

The plots in Fig. 4.3a, b show the energy consumption for the two configurations of the FFT engine with dynamically adaptable pipelines at the clock frequency margined for the worst-case process corner and considering the two typical 5% and 10% voltage margins. Throughputs of several hundreds of MS/s or more are achieved at above-threshold voltages, whereas near-threshold operation takes place for throughputs of several MS/s to a few hundreds of MS/s. Lower throughputs in the order of a few MS/s and lower are achieved at sub-threshold voltages.

As expected from Sect. 2.3, the deep configuration is more energy optimal than the shallow one at throughput targets on the higher end of the range. In detail, the deep configuration is more energy-efficient throughputs of 3.6 GS/s and above for a 5% voltage margin (3.8 GS/s for a 10% voltage margin). This translates into a clock frequency of 450 MHz and above for a 5% voltage margin (472 MHz for a 10% voltage margin), i.e., a supply voltage of 0.84 V (0.88 V). Accordingly, transistors operate in the above-threshold region under throughput targets that favor the deep

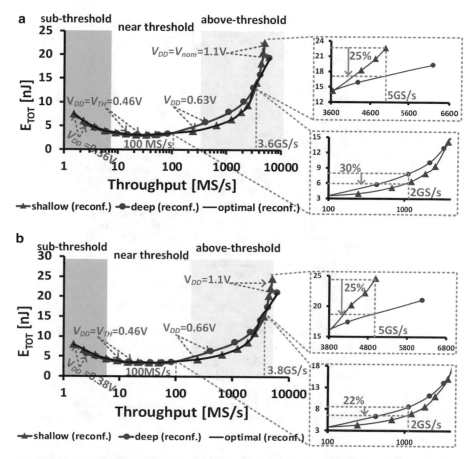

**Fig. 4.3** FFT accelerator energy versus throughput in shallow and deep configuration of a dynamically adaptable pipeline. The clock frequency was set according to the worst-case process corner and voltage corner assuming (**a**) 5% supply voltage margin and (**b**) 10% voltage margin. The optimal configuration is chosen as the most energy-efficient microarchitectural option for each throughput target (i.e., the lowest of the two curves pertaining to the two configurations) [9, 15]

configuration. In this region, the energy gain of the deep configuration over the shallow one is up to 25%, as shown in the inset of Fig. 4.3a, b. The resulting maximum throughput at nominal voltage is 6.3 GS/s. The shallow configuration is more energy-optimal at lower throughputs ranging from 100 MS/s to 3.6 GS/s (3.8 GS/s) for a 5% (10%) voltage margin, and the energy improvement over the shallow configuration is up to 30%, as shown in the inset in Fig. 4.3a, b. In other words, the shallow configuration is more advantageous at near-threshold voltages. At throughputs below 10 MS/s, transistors operate in the sub-threshold region and the leakage energy becomes dominant in all configurations. According to Fig. 2.5b, this leads to very similar energy in both configurations.

Figure 4.4a shows the energy versus $V_{DD}$ and presents the comparison of the two configurations at iso-voltage, as well as for the optimal configuration that corre-

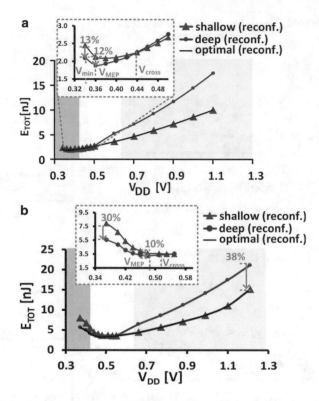

**Fig. 4.4** FFT accelerator energy versus $V_{DD}$ at (**a**) nominal maximum frequency for shallow and deep configurations, (**b**) frequency margined for worst-case process corner and 10% $V_{DD}$ margin

sponds to the lowest energy between the two for each voltage, as enabled by the run-time microarchitectural reconfiguration. From Fig. 4.4a, the shallow configuration is energy optimal at $V_{DD}$ larger than $V_{cross} = 0.44$V as qualitatively expected from Fig. 2.5a, whereas the deep configuration exhibits lower energy at lower voltages. Under a non-margined design, the energy gain at the minimum voltage $V_{min} = 0.34$V, the minimum-energy voltage $V_{MEP} = 0.36$V and the nominal voltage $V_{nom} = 1.1$V are 13%, 12%, and 38%, respectively. When the frequency is margined to account for the worst-case process corner and 10% voltage margin, $V_{cross}$ increases to 0.51 V and the energy savings at $V_{min} = 0.36$V, $V_{MEP} = 0.48$V, and $V_{nom} = 1.1$V become 30%, 10%, and 38% as shown in Fig. 4.4b.

### 4.1.3  Impact of Variations and Comparison of Measurement Results Across Multiple Dice

The above measurements were repeated on 18 dice, and the resulting distribution of the clock frequency at 0.5 V, 0.6 V, and 0.8 V is shown in Fig. 4.5a. From this figure, the average ratio between the clock frequency of the deep and the shallow configuration is 1.7–1.8×. This is well in line with the value of 1.9× expected from simulations, which is expectedly lower than the ideal factor of 2× for the reasons discussed in Sect. 4.1.1. As expected, the clock frequency variability increases when $V_{DD}$ is reduced and approaches the threshold voltage ($V_{TH} = 0.46$ V at the nominal voltage) and is respectively 12% and 8% for deep and shallow configuration at $V_{DD} = 0.5$ V.

The ratio of the clock frequency variability in the deep and shallow configuration is 1.5× and is expectedly close to the reciprocal of the square root of the logic depth ratio (i.e., $\sqrt{2}$), due to the variation averaging effect across cascaded logic gates [5–8].

The histogram of the voltage and the energy at the minimum energy point is plotted in Fig. 4.5b, c. From these figures, the MEP of the deep (shallow) configuration lies in the 0.48–0.51 V (0.53–0.55 V) voltage range, and hence in the near-threshold region. Figure 4.5c shows that the deep configuration at the MEP voltage offers 9% energy reduction on average and 15% in the best case across the measured dice. Comparison of Fig. 4.5b, d reveals that $V_{cross}$ always lies in the near-threshold region and is placed at the right of the MEP of the deep configuration, whereas it is close to the MEP of the shallow configuration. From Fig. 4.5e, the maximum energy reduction is achieved in the above-threshold region, and certainly at the right of $V_{cross}$ (see Figs. 4.3 and 4.4) and the minimum energy point (see Fig. 4.5e). Overall, Fig. 4.5b–e show that the MEP voltage, $V_{cross}$ and the voltage $V_{max,\,gain}$ at which the energy gain is maximum are highly consistent across different dice.

## 4.1.4   Overhead Due to Microarchitecture Reconfiguration

Microarchitectural reconfigure-ability expectedly comes at an energy, throughput, and area overhead due to the insertion of bypassable registers, compared to a conventional fixed pipeline. To fairly evaluate such overhead, the dynamically adaptable pipelined FFT engine in its shallow configuration was compared with the fixed-pipeline version on the same test chip, as they were both designed to have the same logic depth and timing constraints. The clock network of both the reconfigurable and the fixed pipeline was designed by relying on the clock tree synthesis flow of a well-known commercial EDA tool, using the same target clock skew and slew. A conventional non-reconfigurable clock tree was adopted to assure that measurements highlighted the effect of the microarchitectural reconfiguration only, as intended in this chapter. The ability to reconfigure the clock tree will be discussed in Chap. 5.

With regard to the energy overhead due to reconfiguration, Fig. 4.6a plots the measured energy of both the conventional and the reconfigurable version of the FFT engine at the same throughput, along with the energy penalty entailed by reconfiguration. From this figure, the energy overhead is 5.7% at the maximum throughput (i.e., nominal voltage), 7.9% at the intermediate throughput of 700 MS/s (about 10× lower than maximum, as achieved at 0.6 V). From Fig. 4.6a, the overhead due to reconfiguration across dice has a maximum value of 10.5% at the lowest throughput (achieved at $V_{min} = 0.34$ V), and its overall value averaged across voltages and dice is 6%.

Energy results are consistent as plotted in Fig. 4.6b, which shows the energy at 0.6 V for 18 measured dice. The statistical distribution in this figure shows that the

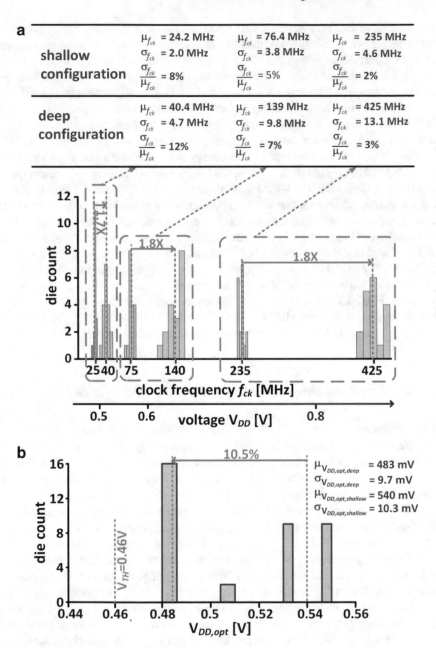

**Fig. 4.5** Histograms showing the statistical distribution across 18 dice of (**a**) the maximum clock frequency $f_{CK}$ for deep and shallow configuration at 0.5 V, 0.6 V, and 0.8 V, (**b**) the energy-optimal voltage $V_{DD, opt}$ at the minimum-energy point, (**c**) the energy at the minimum-energy point, (**d**) the voltage $V_{cross}$ defining the voltage sub-ranges in which one of the two configurations exhibits an energy saving, (**e**) the voltage $V_{max, gain}$ at which the maximum energy gain occurs

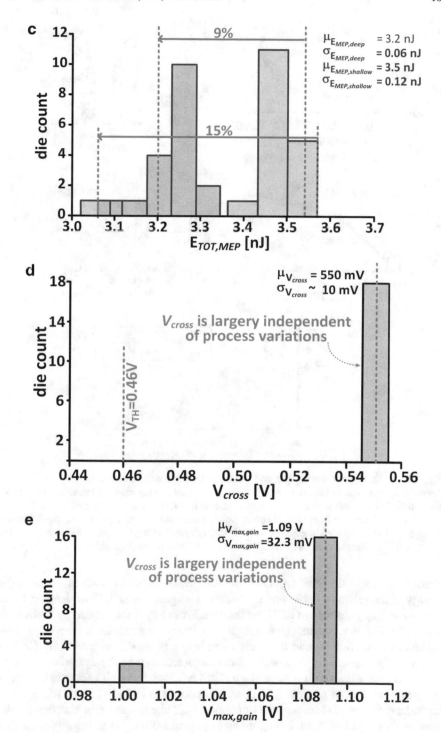

Fig. 4.5  (continued)

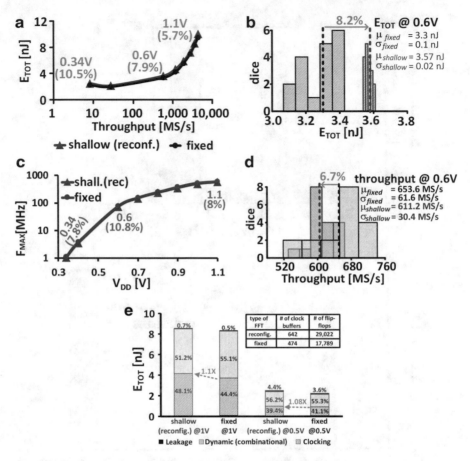

**Fig. 4.6** (**a**) Energy vs. throughput and percentage energy overhead due to reconfiguration at three-pipestage compared to conventional fixed three-pipestage microarchitecture (as indicated in parentheses), (**b**) energy distribution across 18 dice at 0.6 V, (**c**) maximum throughput vs. $V_{DD}$ and percentage performance penalty over conventional fixed microarchitecture due to reconfiguration at iso-voltage (as indicated in parentheses), (**d**) throughput distribution across 18 dice at 0.6 V, (**e**) energy per cycle breakdown compared to conventional fixed microarchitecture

average energy of the fixed (shallow) pipeline design is 3.3 nJ, whereas the average energy of the reconfigurable one in shallow configuration is 3.57 nJ, leading to an average energy overhead of 8.2% across dice. From Fig. 4.6b, the energy variability $\mu/\sigma$ across dice is very small (lower than 3%), confirming that the impact of process variations is negligible, and the energy can hence be predicted accurately for any die. Such low variability is expected from the dominance of the dynamic energy at above- and near-threshold voltages (see Fig. 4.6e), which is well known to be essentially independent of process variations, as opposed to leakage energy [6]. Similar considerations hold at throughputs obtained with above- and near-threshold voltages as shown in Fig. 4.3a, b (i.e., down to several MS/s), due to the dominance of the dynamic energy. At very low throughputs close to the minimum (i.e., very few MS/s), the inevitably higher energy variability due to the more significant leakage energy contribution is irrelevant, as the resulting energy is higher than the minimum, making operation at such design points impractical.

Regarding the throughput degradation due to reconfiguration, simulations at nominal conditions showed 5% maximum throughput degradation at the nominal corner, voltage, and temperature compared to the fixed deep pipeline. This is due to the additional delay of the multiplexers inserted in bypassable registers. The latter is indeed lower than 2FO4 and is responsible for less than 6% of the 33FO4 clock cycle achieved (see Fig. 4.1b). Measurements at room temperature in Fig. 4.6c plot the throughput versus $V_{DD}$ and the penalty due to reconfiguration. From this figure, the performance penalty can be as high as 10.8% at 0.6 V, and it decreases to 6.7% when averaged across the 18 considered dice at the same 0.6 V voltage (see Fig. 4.6d). Overall, the performance penalty averaged across voltages and dice is 6.7%, which accounts for both the additional multiplexer delay and its process variations. As shown in Table 4.1, the area overhead due to reconfiguration is 3.1%, when fairly comparing the dynamically adaptable design with the fixed deep pipeline version.

Figure 4.6e summarizes the measured energy breakdown for the reconfigurable and the fixed microarchitecture. The clocking energy was measured by disabling the FFT logic by reset assertion, while still running the clock. The clocking energy contributes to 48.1% (44.4%) of the total energy budget in the reconfigurable (fixed) microarchitecture. Such relatively large contribution of the clocking energy is mainly due to the large number of latches used in the FIFO embedded in each *commutator* block in Fig. 4.2 [9]. The clocking energy overhead is 10% (8%) at 1 V (0.5 V) and is mainly associated with the extra clock buffers and gaters in the clock network of reconfigurable microarchitecture. The latter ones are needed to selectively enable the clock signal in bypassed registers, as discussed in Sect. 3.8.

As further benefit of microarchitectural reconfiguration, the sensitivity of the energy to $V_{DD}$ around the minimum energy point is significantly reduced. This is shown in Fig. 4.7, which plots the percentage energy increase with respect to the energy at the minimum energy point, when the voltage deviates from its correct voltage by $\Delta V_{MEP}$. The latter summarizes the effect of inaccuracies and discretization in the voltage generation circuitry. Due to the exponential leakage energy increase at voltages lower than the minimum energy point [10], substantial energy increase is experienced even under small voltage deviations in fixed microarchitectures. In dynamically adaptable pipelines, the selection of the optimal deep configuration around the MEP mitigates the energy degradation by 63% (41%), compared

**Table 4.1** Area overhead due to reconfigurations and area breakdown by type of cell (sequential cells = flip-flop + FIFO latches)

| FFT | No. of sequential cells | No. of combinational cells | Area (mm)$^2$ |
|---|---|---|---|
| Fixed (shallow) | 32k | 150k | 0.270 |
| Fixed (deep)[a] | 43.3k | 179k | 0.384 |
| Reconfiguration | 43.3k | 188k | 0.396 |
| Overhead (reconfiguration to fixed (deep)) | 0% | 5% | 3.1% |

[a]Evaluated from P&R

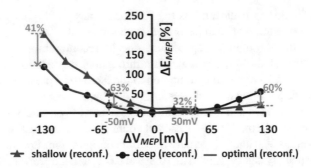

**Fig. 4.7** Percentage energy increase $\Delta E_{\text{MEP}}$ with respect to the MEP energy in the optimal configuration of the reconfigurable microarchitecture versus voltage deviation $\Delta V_{\text{MEP}}$ with respect to the MEP voltage. The adoption of the optimal configuration extends the flat energy region towards the left (right), compared to the single shallow (deep) configuration

**Table 4.2** Comparison of reconfigurable FFT with state-of-the-art FFT engines at minimum energy point

|  | [1] | [2] | [3] | [4] | This work [9] |
|---|---|---|---|---|---|
| Technology | 65nm | 65nm | 180nm | 90nm | 40nm |
| Size | 1024 | 128–2048 | 128–1024 | 256 | 256 |
| Word width | 16 bit | 12 bit | 16 bit | 10 bit | 16 bit |
| Area | 8.5mm$^2$ | 1.37mm$^2$ | 5.5mm$^2$ | 5.1mm$^2$ | 0.53mm$^2$ |
| Design point | CV 0.27V, 30 MHz,240MS/s | CV 0.43V, 10 MHz,80MS/s | RV 0.35V, 10 KHz,NA | CV 0.85V, 300 MHz,2.4GS/s | CV 0.36V, 2.5 MHz,20MS/s |
| Energy/FFT$^a$ | 15.8 nJ | 6.2 nJ | 30 nJ | 12.8 nJ | 1.88 nJ |
| Normalized energy/FFT$^b$ | 2.43 nJ | 5.6 nJ | 6.7 nJ | 10.5 nJ | 1.88 nJ$^a$ |

$^a$All measured at true maximum (non-margined) frequency

$^b$Energy normalized as in: $E_{\text{norm}} = \text{energy} * \dfrac{256}{\text{FFTsize} * \dfrac{\text{tech}}{40\text{nm}} * 0.66 * \dfrac{\text{WL}}{16} + 0.33 * \dfrac{\text{WL}^2}{16}}$

$^c$Energy at margined frequency is 2.6 nJ

to the fixed shallow configuration at $\Delta V_{\text{MEP}}$ equal to $-50$ mV ($-125$ mV) as in Fig. 4.7. When $V_{\text{DD}}$ exceeds the MEP voltage by 50 mV or more, the optimal configuration becomes the shallow one, which reduces energy by 32% (60%) compared to the fixed deep configuration. As a result, the correct choice of the microarchitecture configuration at the MEP mitigates the traditionally significant energy dependence on $V_{\text{DD}}$, relaxing the accuracy requirement of the supply regulation circuitry to track the minimum energy point.

From Table 4.2, the above FFT engine with microarchitectural reconfiguration achieves the lowest energy per FFT computation of 1.88 nJ, under the energy-optimal configuration and minimum energy point voltage $V_{\text{MEP}} = 0.36$ V, compared to prior art [1–4]. For the sake of fairness, the energy consumption of the FFT testchip in this work is evaluated at its true (non-margined) maximum frequency, as was assumed in [1–4]. To make the comparison fair and independent of the technology and the specific FFT parameters (e.g., number of points, word length), the energy was normalized according to the popular figure of merit in [1] under the adopted FFT parameters (256 point, 16-bit input, and 32-bit output bitwidth). From Table 4.2, the dynamically adaptable pipeline design achieves an energy improvement of 1.3–5.6× compared to [1–4].


## 4.2   Finite Impulse Response Filter (FIR) and Fixed-Point Multiplier

As further test of vehicles to gain an insight into the benefits and the overheads of dynamically adaptable pipelines, a fixed-point multiplier and an FIR filter were also designed in the same 40-nm technology. The fixed-point multiplier is representative of a linear pipeline (see Sect. 2.2), and the FIR filter is a feedforward pipeline. The test vehicles in shallow and deep configuration were again verified against the original shallow and deep fixed microarchitecture through dynamic verification.

The fixed-point multiplier is based on a Booth-recoded Wallace tree architecture with 32-bit output and 16-bit inputs. From a design standpoint, the multiplier was preliminarily pipelined and retimed (Steps 1–2 in Fig. 3.1a) to obtain a logic depth of 30FO4, i.e., a clock cycle of 0.9 ns. By applying Steps 3 and 4 in Fig. 3.1a, the design was made reconfigurable at the pipeline level. The clock cycle in the shallow configuration was found to be 62FO4, which is expectedly close to the ideal value of twice the above clock cycle of the initial deep fixed microarchitecture. The fourth-order FIR filter has 16-bit input and 32-bit output and is based on fixed-point arithmetic. The filter was preliminarily pipelined and retimed to obtain a logic depth of 33FO4 (i.e., clock cycle of 1 ns) through Steps 1–2 in Fig. 3.1a. Then, the FIR filter microarchitecture was made reconfigurable through Steps 3–4, leading to an approximately doubled pipeline depth of 60FO4 in the shallow configuration.

The simulated energy per multiplication of the multiplier and FIR filter is plotted versus throughput in Fig. 4.8a, b, where the energy-optimal curve in black is simply obtained by selecting the configuration having the lowest energy at each throughput. Regarding the multiplier, Fig. 4.8a shows that the deep (shallow) configuration is more energy efficient at high (moderate) throughputs above 80MOPS, as expected from the design considerations in Sect. 2.3. The energy savings of the deep (shallow) configuration at high (moderate) throughputs over the fixed shallow (deep) microarchitecture counterpart is up to 1.25× (1.22×). The ability to reconfigure the microarchitecture for better energy efficiency across a wide voltage range is obtained at the cost of an area (timing) overhead below 4% (9%), compared to the original fixed microarchitecture targeting a 30FO4 clock cycle. The energy overhead ranges from 6 to 11%, when comparing the reconfigurable and fixed counterparts, in both shallow and deep configurations.

The detailed gate count and energy breakdown is provided in Table 4.3. From this table, half of the flip-flops have been made bypassable as expected.

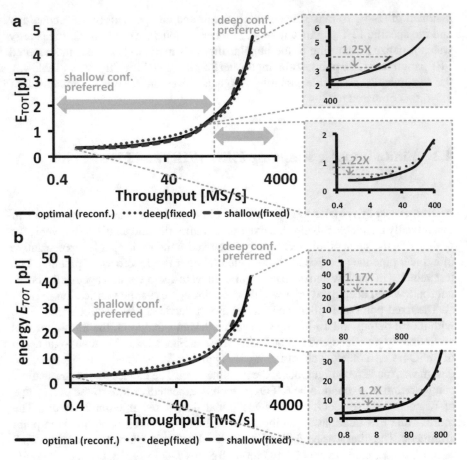

**Fig. 4.8** Energy versus throughput in deep and shallow configuration of (**a**) 16-bit fixed-point multiplier and (**b**) fourth-order 16-bit FIR filter (post-layout simulations)

The energy associated with the combinational logic is larger than the clocking energy, the percentage of the gate count used for sequential elements not being particularly high (i.e., 5–10%).

With regard to the FIR filter, Fig. 4.8b shows that the deep (shallow) configuration exhibits a 1.17× (1.2×) energy reduction compared to the fixed shallow (deep) architecture counterpart, as relevant to high (moderate) throughput targets above (below) 400 MS/s. Such energy reduction is slightly lower than the multiplier and the FFT since the flip-flops account for a very large percentage (i.e., 60%) of the gate count. From Table 4.3, the area overhead due to microarchitectural reconfiguration (i.e., additional multiplexers) is 10%, compared to the fixed deep pipeline counterpart. Again, this larger penalty is due to the dominant presence of flip-flops as a percentage of the overall gate count. The energy overhead ranges from 6 to 11%, similarly to the multiplier. The throughput degradation due to the timing overhead associated with reconfiguration ranges from 4% to 9%.

**Table 4.3** Gate count and energy breakdown in multiplier and FIR filter test vehicles

| | Multiplier | | | | FIR filter | | | |
|---|---|---|---|---|---|---|---|---|
| | Fix (D) | Rec (D) | Fix (S) | Rec (S) | Fix (D) | Rec (D) | Fix (S) | Rec (S) |
| Gate count (combinational) | 1.9 k | 2 k | 1.9 k | 2 k | 6.8 k | 8 k | 6.8 k | 8 k |
| Gate count (sequential) | 200 | 200 | 90 | 200 | 2 k | 2 k | 0.8 k | 2 k |
| # Bypassable flip-flops | – | 105 | – | 105 | – | 1.2 k | – | 1.2 k |
| Energy of combinational logic @ 1.1 V | 2.9 pJ | 3 pJ | 2.9 pJ | 3 pJ | 8.2 pJ | 10 pJ | 8.2 pJ | 10 pJ |
| Energy of sequential logic @ 1.1 V | 2 pJ | 2 pJ | 1.1 pJ | 1.3 pJ | 35 pJ | 35 pJ | 19 pJ | 21 pJ |
| Total energy @ 1.1 V | 4.9 pJ | 5 pJ | 3 pJ | 3.2 pJ | 43 pJ | 45 pJ | 27 pJ | 31 pJ |
| Area overhead | – | 4% | – | – | – | 10% | – | – |
| Energy overhead | – | 6–10% | – | 6–10% | – | 6–10% | – | 5.7–10% |
| Throughput overhead | – | 5–8% | – | 6–9% | – | 5–8% | – | 4–9% |

*D* deep, *S* shallow, *fix* fixed pipeline, *rec* reconfigurable microarchitecture

As a general observation, the timing overhead of reconfiguration is generally due to the additional multiplexer delay, which is typically around 2FO4. Such overhead may be reduced when the critical paths of the individual pipestages being merged are never excited simultaneously, i.e., the critical path delay in the shallow configuration is lower than the sum of the critical paths of the pipestages being merged.

## 4.3  Reconfigurable Thread-Level in ARM Cortex Microcontroller and SRAM Row Aggregation

Reconfiguration was also inserted in an ARM Cortex-M0 processor in 40 nm, as shown in Fig. 4.9. The total core area of 0.3 mm² includes two 8-kB SRAM arrays for program and data memory (0.032 mm² each) based a commercial six-transistor bitcell, and the testing harness (0.15 mm²). The microcontroller and the memory interact with the reconfigurable memory through an AHB bus, as shown in Fig. 4.10.

In Fig. 4.10, the *translator logic* governs the microcontroller–memory interaction through an AHB bus. The *translator logic* has no effect on such interaction when the system is operating in conventional single-thread mode P, as selected by setting the microcontroller dual-thread reconfiguration signal *DT* to 0. As discussed in Sect. 2.7, the translator logic flips the MSB of the program, and data memory addresses every cycle under the throughput-enhanced dual-thread mode P+, when *DT* is set to 1. This enables alternate access of each thread to the relevant portion of the memory space, which is split into two halves with same size for simplicity and with no loss of generality. Regarding the memory, conventional operation in M mode takes place when the *ROWAGG* configuration signal is set to 0, as discussed in Sect. 2.9. On the

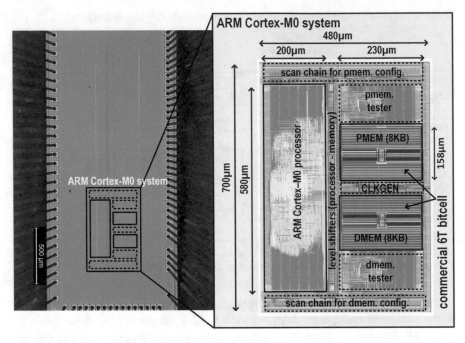

**Fig. 4.9** Testchip of ARM Cortex-M0 microcontroller with dynamically adaptable pipeline and related SRAM memory for data and instructions in 40 nm [16]

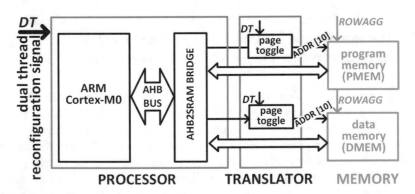

**Fig. 4.10** Reconfigurable ARM Cortex-M0 system architecture [16]

other hand, the M+ mode with shorter access time is triggered when the *ROWAGG* configuration signal is set to 1. Accordingly, four possible system modes are available. Level shifters are inserted between memory and processor to allow independent voltage scaling and co-optimization of their reconfigurable architecture.

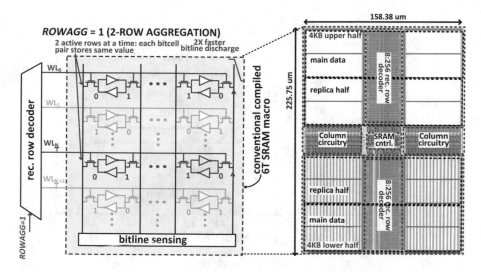

**Fig. 4.11** Architecture and layout of 8-kB SRAM with reconfiguration signal *ROWAGG* [16]

## 4.3.1   Reconfigurable Memory with Selective Row Aggregation

To selectively speed up the read access, memory row aggregation was introduced in M+ mode to double the effective six-transistor bitcell read current being drawn from the bitline capacitance, and hence accelerate its discharge as shown in Fig. 4.11 and discussed in Sect. 2.9. From the latter section, the simultaneous activation of two rows and wordlines is selectively enabled by simply replacing the conventionally inverted address MSB in the second half bank (addresses with MSB = 1) by its XOR with the *ROWAGG* signal. More in detail, under *ROWAGG* = 0 the MSB in the second half bank is simply propagated to the corresponding wordline as in conventional decoders, thus triggering the M mode. When *ROWAGG* = 1, the MSB in the second half bank is inverted and hence made equal to the corresponding row in the first half bank, thus simultaneously activating the bitcells in the two corresponding rows. The wordline pulsewidth is adjusted to shorten the allotted bitline discharge time, to correspondingly reduce the read access time. These slight modifications do not alter the compiled memory array or any other part of the periphery (see Fig. 4.11). Also, the only required modification consists of the insertion of an XOR gate in the row decoder. Accordingly, this approach represents an effective drop-in solution to introduce array reconfiguration that is minimally invasive from a design point of view and allows full reuse of existing compiled memories.

Interestingly, from Fig. 4.12a, b the bitline discharge time variability is reduced by ~25%, thanks to the averaging of the currents delivered by the two simultaneously activated bitcells. This translates into 3× improvement in the bitline discharge time value margined by 6σ, in M+ mode at both $V_{nom} = 1.1$ V and $V_{min} = 0.6$ V. This 3× improvement incorporates the 2× improvement at nominal conditions without considering variations and an additional 1.5× improvement thanks to reduction in the related variability.

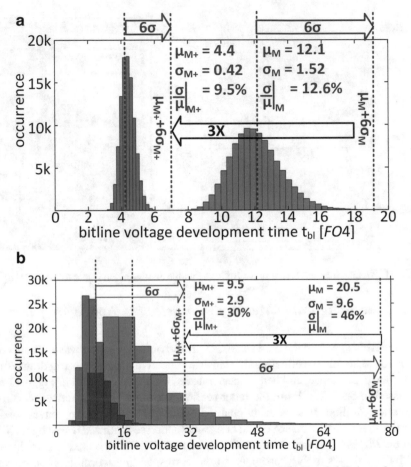

**Fig. 4.12** Monte Carlo simulation in 40 nm at (**a**) $V_{DD}$=1.1 V and (**b**) $V_{DD}$=0.6 V shows 3× speed-up in the bitline discharge time (evaluated at 6-$\sigma$ design margin, 100,000 runs) [16]

When the bitline voltage development be a significant fraction of the overall read access time, the above speed-up via row aggregation in M+ mode translates into a 1.4× to 1.85× access time reduction at 0.6–1.1 V, as shown in Fig. 4.13. Such speed-up is achieved at small energy penalty (1–4%, from Fig. 4.14), compared to the same array without reconfiguration. Such energy overhead is due to the extra word-line that is activated when the SRAM is operating in row-aggregated mode M+, which represents a minor contribution compared to the total SRAM energy consumption. The speed-up in M+ over M mode expectedly improves at lower $V_{DD}$, as the bitline discharge time variability (and hence the usual 6-$\sigma$ margin) becomes larger at low $V_{DD}$, thus emphasizing the above benefit of current averaging. The memory supply voltage is lower bounded by the minimum voltage $V_{min}$ equal to 0.6 V, below which bitcells start failing. Also, Fig. 4.13 shows that write is faster than read; hence, the read access time is the actual timing bottleneck in M+ (even more so in M mode).

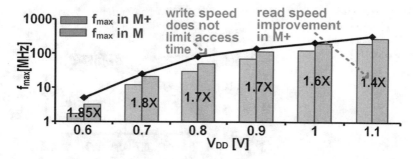

**Fig. 4.13**  Maximum SRAM frequency vs. $V_{DD}$, and speed-up of M+ w.r.t. M mode [16]

**Fig. 4.14**  The reconfigurable SRAM energy per read access vs. $V_{DD}$ in M and M+ mode is nearly the same [16]

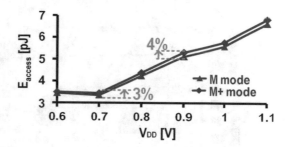

### 4.3.2   ARM Cortex-M0 Microcontroller

In the entire core including both processor and SRAM, the fastest (M+,P+) mode enhances the maximum throughput by 1.8× (Fig. 4.15) and reduces the minimum energy by 1.4× (Fig. 4.16), using matrix multiplication as benchmark in both threads.

The above improvements come at the cost of halved memory capacity, compared to the baseline (M, P) mode. Compared to prior ARM Cortex-M0 research prototypes (see Table 4.4), the proposed core has the lowest minimum energy of 7.4 pJ/cycle in (M+, P+) mode, 8.64 pJ/cycle in (M, P) mode, and the highest maximum throughput, thanks to reconfiguration. The (M+, P+) configuration boosts performance and lowers energy at MEP, whereas (M, P) has lower energy from nominal voltage down to 0.6 V. Compared to a conventional single core with non-reconfigurable processor and SRAM, the above advantages come at the expense of a slight performance degradation. This is due to 6% (4%) processor (memory) timing overhead.

Overall, reconfiguration enables the execution of up to two simultaneous threads and 1.8× throughput increase at only 16.4% larger area, compared to a traditional single core. This is because time interleaving only increases the number of sequential logic gates, while keeping the combinational ones constant (see Sect. 2.6). Similarly, SRAM memory reconfiguration requires only the addition of an XOR gate in the row decoder, which results in zero area overhead. At the same time, reconfiguration significantly reduces the energy at the minimum energy point.

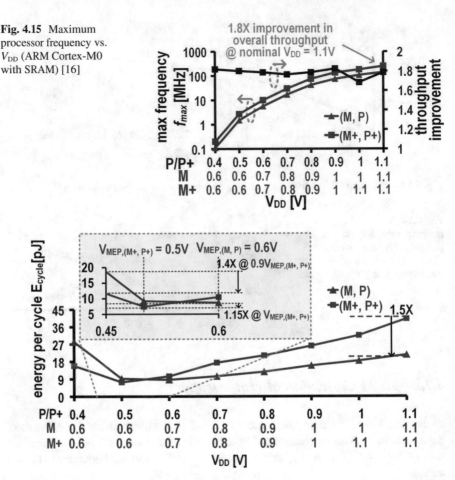

**Fig. 4.15** Maximum processor frequency vs. $V_{DD}$ (ARM Cortex-M0 with SRAM) [16]

**Fig. 4.16** Energy per thread at various configurations of a complete core including an ARM Cortex-M0 and two 8-kB SRAM banks (designed with 10% voltage margin). Matrix multiplication is adopted as a computationally intensive benchmark [16]

In other words, reconfiguration allows to significantly extend the throughput range and the minimum energy beyond allowed by voltage scaling, surpassing the capabilities of conventional voltage scaling as discussed in Chap. 2.

## 4.4   Conclusion

In this chapter, several cases studies of dynamically adaptable microarchitectures at the pipestage and thread level were analyzed through testchip measurements and post-layout simulations. Benchmarks include FFT, FIR, fixed point multiplier, and ARM Cortex-M0 processor including reconfigurable SRAM. Across benchmarks,

**Table 4.4** Performance summary, comparison with state-of-the-art (top, best performance in bold), and overhead of reconfiguration (bottom) [16]

| | Technology | Processor | Area (area/$F^2$) | Gate count (kgates) | Memory size | $V_{DD}$ range | Frequency range | Min. energy/cycle/thread (frequency) | Max. throughput improvement beyond $V_{DD}$ scaling | Min. energy improvement beyond $V_{DD}$ scaling |
|---|---|---|---|---|---|---|---|---|---|---|
| This work (M+,P+) (M,P) | 40nm | ARM Cortex-M0 | **0.16mm²** ($10^6F^2$) | 34 | 16kB | 0.4V–1.1V | 200kHz–250MHz | 7.4 pJ @0.5 V (2.7MHz) | 1.8× | 1.36× |
| | | | | | | | 100kHz–140MHz | 8.64pJ @0.6V (5.56MHz) | 1× | 1× |
| [11] VLSI Symp. '17 | 65nm | ARM Cortex-M0+ | 3.76mm² ($57 \cdot 10^9 F^2$) | 67 | 12kB+2kB ROM | 0.3V–0.8V | 20kHz–50MHz | 12.9pJ @0.35V (174kHz) | 1× | 1× |
| [12] ISSCC '15 | 65nm | ARM Cortex-M0+ | 3.76mm² ($57 \cdot 10^9 F^2$) | 30 | 24kB | 0.25V–1.2V | 20kHz–100MHz | 11.7pJ @0.35V (750kHz) | 1× | 1× |
| [13] JSSC '17 | 40nm | ARM Cortex-M0 | 0.39mm² ($2.4 \cdot 10^8 F^2$) | 36 | 64kB | 0.2V–0.5V | 0.8MHz–50MHz | 43.2pJ @0.37V (15MHz) | 1× | 1× |
| [14] ISSCC '12 | 180nm | ARM Cortex-M0 | 1.7mm² ($5.3 \cdot 10^7 F^2$) | 35 | 3kB | 0.6V | 160kHz–330kHz | 17.2pJ @0.26V (160kHz) | 1× | 1× |

OVERHEAD DUE TO RECONFIGURATION

Processor — MUX — MEM — clk DT — CLK GATER — EXTRA REG — RECONFIG. DECODER — clk DT

| | Gate count | Overhead | Cycle time or delay | Overhead |
|---|---|---|---|---|
| Baseline processor | 25kgates | – | 100$FO4$ (53$FO4$) | – |
| Extra MUXes | 2kgates | 8% | 3$FO4$ (3$FO4$) | 3% (6%) |
| Extra registers | 2kgates | 8% | – | – |
| Extra clock gaters | 0.1kgates | 0.4% | – | – |
| Baseline SRAM | 0.036mm² | – | 125$FO4$ | – |
| Additional decoder reconfig. for row aggregation | 0mm² | 0% | 0 | 0% |

AREA OVERHEAD — TIMING OVERHEAD FOR SINGLE (DUAL) THREAD

AREA BREAKDOWN processor: others (30%), seq. (32%), comb (38%)

memory: per. (34%), bitcells (66%)

dynamically adaptable pipelines at the pipeline level were found to yield a minimum energy and maximum throughput benefit of up to 38% and 80%, with a maximum area overhead of 9%. Thread-level reconfiguration in the ARM Cortex-M0 system with reconfigurable memory was found to yield a minimum energy and maximum throughput improvement by 40% and 80%, respectively, with area overhead of 16.4%. Thanks to the design methodologies developed in Chap. 3, such enhancements beyond the voltage scaling capabilities were obtained at negligible design effort, thanks to full design automation.

# References

1. D. Jeon, M. Seok, C. Chakrabarti, D. Blaauw, D. Sylvester, A super-pipelined energy efficient subthreshold 240 MS/s FFT core in 65 nm CMOS. IEEE J. Solid State Circuits **47**(1) (2012)
2. C.H. Yang, T.H. Yu, D. Markovic, Power and area minimization of reconfigurable FFT processors: A 3GPP-LTE example. IEEE J. Solid State Circuits **47**(3), 757–768 (2012)
3. A. Wang, A. Chandrakasan, A 180-mV subthreshold FFT processor using a minimum energy design methodology. IEEE J. Solid State Circuits **40**(1), 310–319 (2005)
4. Y. Chen, Y.W. Lin, Y.C. Tsao, C.Y. Lee, A 2.4-G sample/s DVFS FFT processor for MIMO OFDM communication systems. IEEE J. Solid State Circuits **43**(5), 1260–1273 (2008)
5. M. Alioto, G. Scotti, A. Trifiletti, A novel framework to estimate the path delay variability on the back of an envelope via the fan-out-of-4 metric. IEEE Trans. Circuits Syst. Pt. I **64**(8), 2073–2085 (2017)

6. M. Alioto, G. Palumbo, M. Pennisi, Understanding the effect of process variations on the delay of static and domino logic. IEEE Trans. VLSI Syst. **18**(5), 697–710 (2010)
7. M. Eisele, J. Berthold, D. Schmitt-Landsiedel, R. Mahnkopf, The impact of intra-die device parameter variations on path delays and on the design for yield of low voltage digital circuits. IEEE Trans. VLSI Syst. **5**(4), 360–368 (1997)
8. K. Bowman, S. Duvall, J. Meindl, Impact of die-to-die and within-die parameter fluctuations on the maximum clock frequency distribution for Gigascale integration. IEEE J. Solid State Circuits **37**(2), 183–190 (2002)
9. S. Jain, L. Lin, M. Alioto, Dynamically adaptable pipeline for energy-efficient microarchitectures under wide voltage scaling. IEEE J. Solid State Circuits **53**(2), 632–641 (2018)
10. M. Alioto, Ultra-low power VLSI circuit design demystified and explained: A tutorial. IEEE Trans. Circuits Syst. Pt. I **59**(1), 3–29 (2012)
11. J. Myers, A. Savanth, P. Prabhat, S. Yang, R. Gaddh, S.O. Toh, et al., A 12.4pJ/cycle sub-threshold, 16pJ/cycle near-threshold ARM cortex-M0+ with autonomous SRPG/DVFS and temperature tracking clocks, in *2017 Symposium on VLSI Circuits, Kyoto, Japan*, (2017)
12. J. Myers, A. Savanth, D. Howard, et al., An 80nW retention 11.7pJ/cycle active subthreshold ARM cortex-M0+ subsystem in 65nm CMOS for WSN applications, in *IEEE ISSCC Digest of Technical Papers, San Francisco (CA)*, (2015), pp. 144–146
13. H. Reyserhove, W. Dehaene, A differential transmission gate design flow for minimum energy sub-10-pJ/cycle ARM cortex-M0 MCUs. IEEE J. Solid State Circuits **52**(7), 1904–1914 (2017)
14. Y. Lee, G. Kim, S. Bang, Y. Kim, I. Lee, P. Dutta, et al., A modular 1mm³ die-stacked sensing platform with optical communication and multi-modal harvesting, in *IEEE ISSCC Digest of Technical Papers, San Francisco (CA)*, (2012)
15. S. Jain, L. Lin, M. Alioto, Automated design of reconfigurable microarchitectures for accelerators under wide voltage scaling, in *IEEE Transactions on Very Large Scale Integration Systems* [in print]
16. S. Jain, L. Lin, M. Alioto, Drop-in energy-performance range extension in microcontrollers beyond VDD scaling, in *2019 IEEE Asian Solid-State Circuits Conference, Macau*, (2019), pp. 125–128

# Chapter 5
# Reconfigurable Clock Networks, Automated Design Flows, Run-Time Optimization, and Case Study

**Abstract** This chapter introduces clock network reconfiguration for wide adaptation from nominal voltage down to deep sub-threshold voltages. Reconfiguration resolves the conflicting repeater insertion requirements at different voltages, in conventional static clock networks. In reconfigurable clock networks, the number of repeater levels is dynamically adapted to the supply voltage to ultimately mitigate the clock skew degradation across a wide voltage range. At nominal voltage, the number of repeater levels is adjusted to the highest value to mitigate the important clock skew contribution of wire delays. At lower voltages, the number of repeaters is progressively lowered to mitigate the increasingly dominant clock skew contribution of repeaters.

**Keywords** Reconfigurable clock network · Wire delay · Gate delay · Bypassable repeater · Clock repeater · Clock tree · Hold margin · Robustness against hold violations · Timing violations · Minimum operating voltage $V_{min}$ · Clock skew · Launching register · Capturing register · Shallow clock network · Deep clock network · Monte Carlo simulations · Clock skew standard deviation · Histogram · DIBL effect · Clock distribution · Clock signal · Boostable clock repeater · Clock root · Clock sink · Clock tree leaves · Clock gater · Dummy clock gater · Bypassable clock gater · Clock tree synthesis · Automated clock tree design · Level balance principle · DVFS · Fast Fourier Transform (FFT) · Clock path replica · Clock skew measurement · Time-to-digital conversion · Vernier delay line · Above-threshold region · Near-threshold region · Sub-threshold region

This chapter introduces clock network reconfiguration for wide adaptation from nominal voltage down to deep sub-threshold voltages. Reconfiguration resolves the conflicting repeater insertion requirements at different voltages, in conventional static clock networks. In reconfigurable clock networks, the number of repeater levels is dynamically adapted to the supply voltage to ultimately mitigate the clock skew degradation across a wide voltage range. At nominal voltage, the number of repeater levels is adjusted to the highest value to mitigate the important clock skew contribution of wire delays. At lower voltages, the number of repeaters is progressively lowered to mitigate the increasingly dominant clock skew contribution of repeaters.

© Springer Nature Switzerland AG 2020
S. Jain et al., *Adaptive Digital Circuits for Power-Performance Range beyond Wide Voltage Scaling*, https://doi.org/10.1007/978-3-030-38796-9_5

The fundamental design tradeoffs in clock networks under wide voltage scaling are first reviewed. Reconfiguration is then introduced in terms of principle and circuit techniques. Automated design flows for clock trees are discussed for seamless integration with wide voltage scaling and validated through a case study and silicon measurement results.

## 5.1   Impact of Clock Network Topology on Clock Skew, Performance, and Hold Margin under Wide Voltage Scaling

In digital sub-systems, the clock network design is heavily affected by the adopted supply voltage [1–5]. In other words, the clock network topology achieving a given clock skew at nearly minimal clocking energy highly depends on the supply voltage. Accordingly, the adoption of a conventional static (fixed) clock network is largely sub-optimal under wide voltage scaling [7–13].

In the following, the impact of clock skew on performance and robustness against hold violations is reviewed in Sect. 5.1.1, whereas the impact of the supply voltage on the clock network requirements is discussed in Sect. 5.1.2. Prior art in the design of clock network for low or wide voltage range is presented in Sect. 5.1.3.

### 5.1.1   Impact of Clock Skew on Performance, Robustness Against Hold Violations, and Energy Efficiency

The clock skew quantifies the non-ideality of the clock network in terms of non-simultaneous distribution of the clock across different registers. Since timing constraints in Sect. 2.1 are defined only in pairs of registers at the beginning and end of a pipestage, let us again consider its basic structure in Fig. 2.1a. The clock skew $t_{SKEW}$ in a pipestage is defined as the difference of the clock edge arrival time at the capturing register and the launching register

$$t_{SKEW} = t_{capturing} - t_{launching}.$$

(5.1)

In Eq. (5.1), clock skew is ideally zero in a clock network with perfectly balanced delays across all paths (from the root to all sink nodes in the clock network, i.e., all register endpoints). In actuality, skew is non-zero due to the different delays through different branches of the clock network, as shown in Fig. 5.1. In realistic designs with well-balanced paths, the relative delay is mostly defined by within-die process variations (i.e., mismatch), and the clock skew sign is random and unpredictable.

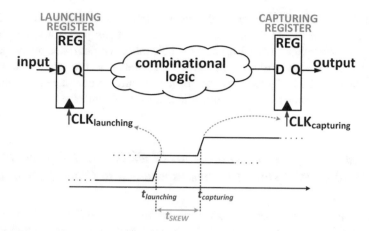

**Fig. 5.1** Definition of clock skew $t_{SKEW}$ in a pipestage. By definition, its sign is positive if the active clock edge arrives first in the launching register and then in the capturing register

The clock skew is well known to affect both performance and robustness against hold violations [4, 5]. Indeed, the clock skew is integral part of the clock uncertainty $T_{cu}$ in the minimum clock cycle in Eq. (2.1), which can be rewritten as follows to highlight the role of the skew [4]:

$$T_{CK,min} = \tau_{comb,PD} + \left(T_{setup} + T_{cq}\right) - t_{SKEW} \tag{5.2}$$

where $\tau_{comb, PD}$ is the combinational propagation delay, $(T_{setup} + T_{cq})$ is the register timing overhead due to the setup time $T_{setup}$ of the capturing register and the clock-to-output delay $T_{cq}$ of the launching register, and $t_{SKEW}$ is the clock skew between these two registers in Eq. (5.1) (see details in Sect. 2.1). From Eq. (5.2), the presence of (negative) clock skew increases the minimum clock cycle and hence reduces the clock frequency, thus degrading the throughput by the same factor. In a fixed clock network, the worst-case negative clock skew must be budgeted at design time to achieve an adequately large design margin assuring correct functionality of the manufactured dice at the targeted frequency.

At the same time, the clock skew reduces the hold margin in Eq. (5.3), which is defined as the maximum cumulative deviation of timing parameters that still prevent hold timing failures [4, 5]

$$t_{hold\,margin} = T_{comb,CD} + T_{cq} - T_{hold} - t_{SKEW} \tag{5.3}$$

where $T_{comb, CD}$ is the contamination delay of the combinational path (i.e., the minimum combinational delay across all possible inputs, as discussed in Sect. 2.1), and $T_{hold}$ is the hold time of the capturing register. A higher $t_{hold\,margin}$ indicates that the digital circuit is able to withstand a larger cumulative change in the parameters in Eq. (5.3), including the usually dominant timing contribution of the clock skew. Hence, a higher $t_{hold\,margin}$ translates into more robust operation. Such hold margin degradation is routinely compensated at design time by inserting extra hold-fix buffers, at the cost of higher complexity, and hence area and power consumption. As opposed to performance, from Eq. (5.3) the hold margin is degraded by the presence of positive clock skew. Again, in a fixed clock network, the worst-case positive clock skew must be budgeted at design time to assure an adequate hold margin to withstand the expected range of within-die across the manufactured dice.

### 5.1.2   Impact of Supply Voltage on Clock Network Optimization and Clock Skew in Conventional Static Clock Networks

In general, a larger number of clock repeater levels is needed at above-threshold voltages, to reduce the wire length between successive repeaters, so that the wire RC time constant is kept in optimal balance with the repeater delay [14–19]. In particular, the RC time constant needs to be kept small enough to avoid an excessive degradation in the clock slope at the input of the next repeater and hence meet the clock slope requirement [4, 5]. Accordingly, deeper clock networks with a larger number of repeaters from the clock root to the flip-flops are preferred, as they reduce the clock skew induced by process variations at the nominal voltage, as depicted in Fig. 5.2.

At near- and sub-threshold voltages, the gate delay significantly increases and hence dominates over the wire RC time constant. In this case, the clock slope degradation induced by wires is no longer an issue, and the random clock skew is dominated by within-die variations in clock repeaters. In addition, the latter skew contribution further increase at low voltages [20] and with the number of repeater levels [4]. Accordingly, shallow clock networks are preferred at low voltages, so that the dominant skew contribution due to clock repeaters is mitigated.

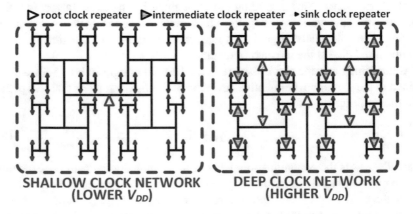

**Fig. 5.2** Shallower clock networks (left) with fewer clock repeaters from root to sink nodes are required at lower voltages. Deeper clock networks are instead more appropriate at higher voltages

From the above considerations, a fixed clock network designed around a specific supply voltage inevitably suffers from clock skew degradation, when experiencing the inevitably large voltage deviations determined by wide voltage scaling. In particular, if the clock network is conventionally designed for above-threshold voltages, the clock skew increases at lower voltages and causes the performance and robustness degradation discussed in the previous subsection [5]. On the one hand, skew-induced performance degradation at lower voltages leads to the need for higher supply voltage to meet the targeted clock frequency, leading to an increase in the energy per operation (see Sect. 2.3). On the other hand, the degraded hold margin requires further mitigation of the effect of within-die variations, which in turn translates into an increase in the minimum supply voltage $V_{min}$ that allows correct operation with no hold time failures [21]. Again, this limits opportunities to reduce energy through aggressive voltage scaling (see details in Sect. 5.6) and increases the energy. This explains why clock skew invariably determines a degradation in the energy efficiency, in addition to performance and robustness degradation. Alternatively, the degraded hold margin can be compensated by increasing the contamination delay in Eq. (5.3), by inserting a larger number of hold-fix buffers [4, 5]. Again, their additional energy contribution degrades the overall energy efficiency, as in the case of operation at higher $V_{min}$. Dual considerations hold for clock networks designed for low voltage, while operating at higher voltages.

As an illustrative example of clock skew-induced performance degradation, Fig. 5.4a plots the random clock skew and slope in a deep (eight-level) and shallow (two-level) clock network in the 40-nm FFT processor testchip in Sect. 4.1. The deep clock network was designed for operation at nominal voltage, whereas the shallow network was designed for operation at the lowest voltage of 0.3 V. The clock skew was evaluated through 10,000-run Monte Carlo simulations of the critical clock path. From Fig. 5.4a, the clock skew substantially increases from 2FO4 to almost 8FO4 delays, becoming slightly more than 20% of the 33$FO4$ clock cycle in deep microarchitecture configuration. From Fig. 5.3b, this clock skew increase expectedly leads to the same performance degradation by more than 20%, under a deep clock network running in the near- and sub-threshold regime. For completeness, the same figure shows that the clock slope remains almost unaltered at very low voltages, as expected from the fact that the wire delay becomes insignificant in the sub-threshold region. Conversely, from Fig. 5.4a the shallow clock network designed at 0.3 V exhibits a clock skew of about 2FO4 at its targeted supply voltage, whereas it rapidly increases to several FO4 delays in the near-threshold region. The resulting percentage performance degradation when up-scaling $V_{DD}$ is expectedly the same from Fig. 5.4b. From the same figure, reliable operation is prohibited at above-threshold voltages. As expected, the clock slope substantially increases to more than 10FO4 at above-threshold voltages, as the inadequately low number of cascaded repeaters makes the intermediate wires much longer and hence slow.

Regarding the clock skew-induced hold margin degradation in the same FFT testchip, Fig. 5.4a shows that the hold margin of the deep network designed at 1.1 V steadily decreases when $V_{DD}$ is reduced. Since it drops below zero for $V_{DD} < 0.6$ V, hold timing failures take place at lower voltages and hence impose

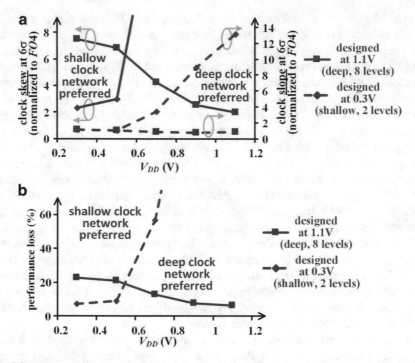

**Fig. 5.3** FFT accelerator as illustrative example of (**a**) clock skew and slope degradation in deep clock network designed for nominal voltage (blue curves), and shallow network designed for 0.3 V (red curves), (**b**) resulting percentage performance loss. The plots are obtained via 10,000-run Monte Carlo simulations in 40 nm

$V_{min} = 0.6\,V$. Alternatively, to prevent such timing failures, more hold-fix buffers need to be inserted to bring the hold margin in Eq. (5.3) back to positive. Figure 5.4b plots the hold-fix buffer count required in the above FFT accelerator to maintain a minimal 3-$\sigma$ hold margin. From this figure, the hold-fix buffer count assuring positive hold margin in the sub-threshold region is 9k gates, resulting in an overall FFT area increase by 1.2%. Regarding the shallow clock network designed at 0.3 V, Fig. 5.4a shows that the hold margin rapidly decreases when $V_{DD}$ is increased, and reliable operation is prohibited at $V_{DD} > 0.5\,V$. From Fig. 5.4b, the necessary hold-fix buffer count to extend reliable operation up to 1.1 V is 156 k gates, accounting for 21.4% FFT area increase. In general, such hold-fix area cost under wide $V_{DD}$ scaling is highly design dependent,[1] difficult to predict, and can represent a quite significant fraction of the overall area [21–25], especially at finer technologies and ampler hold margin targets.

---

[1] The cost of hold fix is determined by several factors such as type and number of critical paths and flip-flops, the clock network size, the nominal skew target.

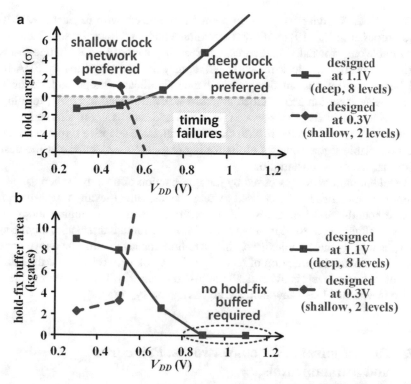

**Fig. 5.4** FFT accelerator as illustrative example of (**a**) 3-σ hold margin degradation at low volt-ages in deep clock network designed for nominal voltage (blue curve) and at higher voltages in shallow network designed for 0.3 V (red curve), (**b**) resulting increase in the area occupied by hold-fix buffers. The plots are obtained via 10,000-run Monte Carlo simulations in 40 nm

### 5.1.3   Prior Art in Clock Networks for Low- or Wide-Voltage Operation

Various approaches were proposed to mitigate the clock skew penalty imposed by static clock networks designed around a fixed voltage. For example, [25, 26] pro-posed moderately deep networks with long-channel and low-threshold voltage buf-fers. Indeed, long-channel voltage buffers avoid the traditional super-linear hold-fix buffer delay increase at low voltages, due to the transistor threshold voltage increase determined by the well-known DIBL effect. The adoption of low transistor thresh-old voltages in hold-fix buffers further mitigates the effect of DIBL and mitigates the delay increase at low voltages. As other approaches, static clock network design methodologies extending the voltage range were investigated in [14–17]. However, such methodologies cannot escape from the fundamentally opposite requirements imposed by low and nominal voltage operation, and hence still suffer from the above-discussed limitations.

Techniques for adaptive point-to-point interconnects with regenerative drivers were proposed in [18, 19] to allow wide voltage scaling. However, such techniques are suitable only for individual wires, whereas they cannot be used in one-to-many interconnects such as clock networks. Also, they are not supported by EDA tools and hence require custom design and hence significant design and verification effort. Delay insertion across clock domains was introduced in [7, 27] to mitigate the inter-domain skew (e.g., between processor and accelerators), although voltage adaption was not considered in each clock domain. Such adaptive signal distribution methods enable tighter control on its timing (e.g., skew reduction), at the cost of additional area and consumption.

As additional challenges posed by the introduction of adaptive schemes, proper tuning strategies need to be devised to keep the additional testing time within reasonable bounds, and hence cost. Also, integration with other tuning loops (e.g., voltage scaling) needs to be handled to minimize the mutual interaction among such loops, as well as the complexity of their run-time management. The above considerations motivate the adoption of reconfigurable clock networks where the number of clock repeater levels is dynamically adjusted along with $V_{DD}$, preventing the otherwise inevitable clock skew increase across a wide $V_{DD}$ range.

## 5.2  Reconfigurable Clock Networks: Principles and Fundamentals

To overcome the limitations of conventional static clock networks, the concept of reconfigurable clock networks is depicted in Fig. 5.5a. In reconfigurable clock networks, each repeater is made bypassable to selectively suppress those that are unnecessary when the supply voltage is reduced, extending the concept of register bypassing in data paths to repeaters in clock paths. In normal mode, each reconfigurable repeater is configured as a conventional buffer. In bypass mode, repeaters are bypassed to merge the previous and subsequent wire, suppressing the corresponding repeater level.

The clock network is configured in its deepest configuration by setting all repeaters in normal mode at the nominal voltage, as shown in Fig. 5.5b. When $V_{DD}$ is scaled down, the number of repeater levels is reduced by progressively bypassing the intermediate bypassable clock repeaters, making the clock network shallower (Fig. 5.5c). The circuit details on bypassable repeaters are discussed in Sect. 5.3.

In the proposed scheme in Fig. 5.5a, the clock root repeater is different from all other repeaters, since its capacitive load experiences the largest increase when bypassing the subsequent repeaters. Indeed, in the worst case where all intermediate repeaters are bypassed, the first root repeater needs to drive the entire wire capacitance of the clock network. Repeaters in the next levels typically need to drive an exponentially lower capacitive load, making the clock root repeater more critical than others. For example, in a balanced H tree, the wire load seen by second-level

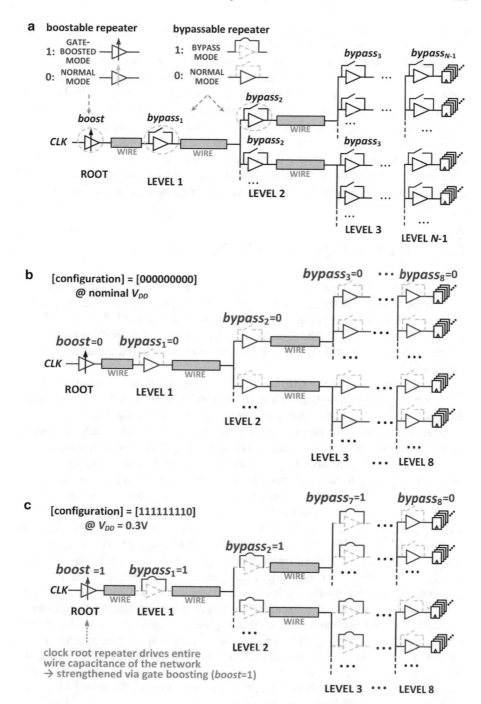

**Fig. 5.5** (**a**) Reconfigurable clock network, (**b**) deepest configuration (000000000), (**c**) shallowest configuration (111111110). The first bit of the configuration is the *boost* signal (0 if clock root is not gate boosted, 1 otherwise), the successive bits are the *bypass* signals (0 in normal mode, 1 in bypassed mode)

(third-level) repeaters is 2× (4×) lower than the clock root, leading to an exponential reduction by a factor of $2^{i-1}$ at the $i$-th level [4].

In the above worst-case load where all intermediate repeaters are bypassed, the strength of the root clock repeater needs to be increased to drive the entire clock network capacitance associated with its wires, as in Fig. 5.5c. It is worth observing that up-scaling its strength in a static manner (e.g., transistor over-sizing at design time) is not really an option, as this would inevitably increase the clocking energy due to the large size and the full activity of the clock signal. Accordingly, it is particularly important to introduce a mechanism that selectively and dynamically increases the strength of the clock root repeater without significantly increasing its energy or size. This is achieved through gate boosting, as discussed in Sect. 5.4.

## 5.3   Bypassable Repeaters and Other Clock Cells

The schematic of the bypassable clock repeater is shown in Fig. 5.6, where the normal/bypass mode is set by the *bypass* signal. When *bypass* = 0, the circuit operates as a conventional CMOS buffer, as enabled by the transistors M1–M4 being ON in Fig. 5.6. When *bypass* = 1, the repeater circuitry M1–M4 is disabled via power gating to turn transistors M5–M8 off, and the repeater is bypassed by turning on the pass transistor M9.

To avoid the threshold voltage loss across the pass transistor M9 in bypass mode, it is interesting to observe that clock repeaters need to be bypassed only when operating at low $V_{DD}$ (e.g., near- and sub-threshold), since a shallower network is targeted at such voltages. Accordingly, the threshold voltage loss is naturally eliminated by simply driving the gate terminal of M9 with the available nominal voltage $V_{DD, nom} = 1.1$ V, as it easily exceeds $V_{DD}$ by more than a threshold voltage. Indeed, as will be shown in Sect. 5.6, clock network reconfiguration takes place at near-threshold voltages and below, which are below the nominal voltage by several hundreds of mV. It is useful to observe that the adoption of a gate voltage $V_{DD, nom}$ in M9 greater than $V_{DD}$ does not cause any dynamic energy penalty, since it is kept constant in bypass mode. As further side benefit, the adoption of $V_{DD, nom} > V_{DD}$ increases the strength of M9, thanks to its inherent gate boosting. At the expected near- and sub-threshold $V_{DD}$ under clock network reconfiguration, the strength of M9 is easily increased by an order of magnitude or more [3]. In turn, this reduces the area overhead due to M9 for a targeted strength, which is typically set to make its equivalent resistance insignificant compared to the neighboring wire resistance. As further benefit, the reduced size of M9 mitigates the parasitic capacitance at the repeater input and output, which in turn defines the clocking energy overhead of bypassable clock repeaters. From layout parasitic extraction, such parasitic capacitance was found to increase the capacitance (i.e., clocking energy) of a conventional repeater with the same strength by at most 11%.

Compared to a conventional repeater with same strength, the bypass-ability in Fig. 5.6 comes at a minor performance and power penalty. As shown in Fig. 5.7,

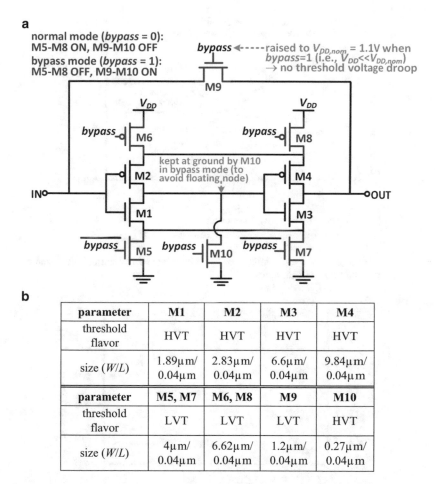

**Fig. 5.6** (a) Schematic of bypassable clock repeater and (b) detailed transistor sizes for the 24×
standard cell

20% delay overhead is experienced by the bypassable repeater compared to the
conventional repeater with the same 24× strength. Such delay increase is nearly
independent of the supply voltage and is due to the stacked transistors M5–M8 and
to the additional parasitic associated with the pass transistor M9. This result is still
valid when variations are included, as Fig. 5.7 shows that the delay variability due
to within-die variations is essentially the same as the conventional repeater at above-
threshold voltages. At near- and sub-threshold voltages, the delay variability of the
bypassable repeater is still close to the conventional one, with a maximum difference of 14%. As will be discussed in Sect. 5.6, the potential robustness benefit
obtained from the clock network reconfiguration is fully retained because the variability increase is mostly negligible. The power overhead of a bypassable buffer is
less than 10%, as expected from the limited additional parasitic capacitance, as
discussed above (see Fig. 5.8).

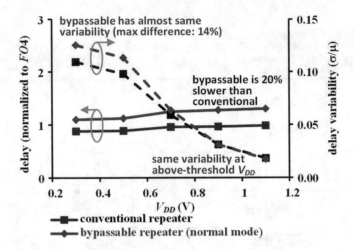

**Fig. 5.7** Simulated delay and its variability of bypassable repeater in normal mode and conventional repeater at the same 24× strength

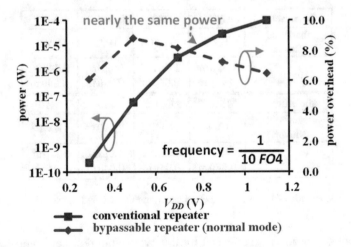

**Fig. 5.8** Power consumption and power overhead of bypassable clock repeater compared to conventional repeater at same 24× strength

The circuit approach in Fig. 5.6 to introduce bypassability in clock repeaters can be immediately extended to any logic gate lying in the clock path. For example, this applies to clock gaters [28], as they are an essential part of clock networks and hence need to have the ability to merge the previous and subsequent wire, as required in bypassable clock networks. A possible implementation is respectively shown in Fig. 5.9a, b in its gate- and transistor-level view. Being in the clock path, the AND gate of the gater in Fig. 5.9a needs to be made bypassable, as achievable through the

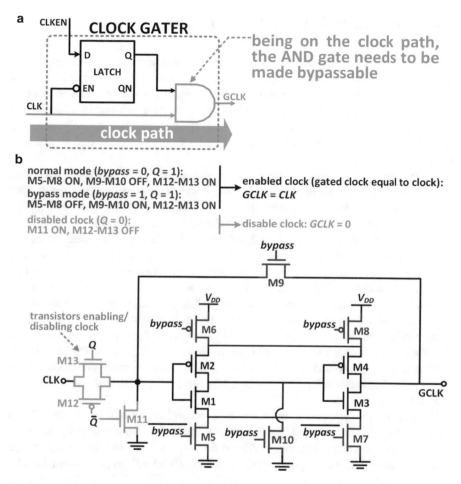

**Fig. 5.9** (**a**) Gate-level view of clock gater enhanced with bypassability via bypassable AND gate, (**b**) possible circuit implementation of bypassable clock gater

same approach as in Fig. 5.6. In the resulting clock gater with bypassable AND gate in Fig. 5.9b, the global clock *CLK* is propagated to the output gated clock *GCLK* when $Q = 1$ via M12-M13, in both normal and bypass modes. On the other hand, the gated clock *GCLK* is inhibited and kept low when $Q = 0$, which switches off M12–M13 and turns on M11, forcing the output gated clock to 0.

To maintain balanced clock path delays across all configurations, the number of levels in every clock path needs to be the same. As discussed in Sect. 5.5, this requires balanced insertion of clock gaters across all clock paths, as achieved through dummy clock gater insertion in paths where intended clock gaters are missing.

## 5.4  Gate-Boostable Clock Root Repeater

As discussed above, the strength of the clock root repeater needs to be selectively and occasionally increased, when most or all clock network levels next to the root are simultaneously bypassed. As shown in Fig. 5.5c, this typically occurs at the lowest end of the considered $V_{DD}$ range and hence in the deep sub-threshold region. In the latter region, the strength of the clock root repeater can be effectively increased through gate boosting, thanks to the exponential strength dependence on the gate voltage [29].

In general, gate boosting voltages can be generated *in situ* through a small and local charge pump or bootstrapping [3, 27–29] and can be then applied to the input of the clock root repeater via a level shifter. This also allows significant testing flexibility for circuit exploration. As shown in the adopted circuit in Fig. 5.10, M1 and M2 are respectively gate-boosted by level shifters LS1 and LS2 in boost mode, i.e., when *boost* = 1 and path 2 is hence activated by M5–M6. In addition, M1 and M2 are low-threshold voltage (LVT) transistors to increase their driving strength at low voltages. As shown by Fig. 5.11, the clock root repeater strength increases by up to four orders of magnitude, when the boosting voltage $\Delta V_{boost}$ is set to the moderately high value of 300 mV at $V_{DD}$ < 400 mV. $\Delta V_{boost}$ represents an effective knob to dynamically and flexibly adjust the driving strength of the clock root repeater, so that it can drive a very wide range of capacitive loads, as required when all intermediate levels are bypassed. On the other hand, operation in normal mode occurs when *boost* = 0, which activates path 1 through M5–M6 in Fig. 5.10, and drives the clock output via G2.

From a design perspective, the clock root repeater strength is set via the G2 and M3–M4 size to drive the first level in normal mode (i.e., conventionally designed under the deepest clock network configuration) and via the M1–M2 and M5–M6 size to drive the total wire capacitance load at $\Delta V_{boost}$ = 300 mV in gate-boosted mode.

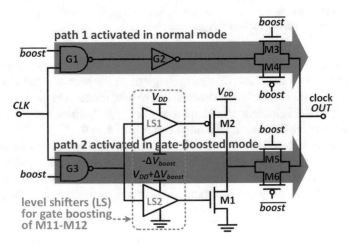

**Fig. 5.10**  Schematic of gate-boosted clock root repeater

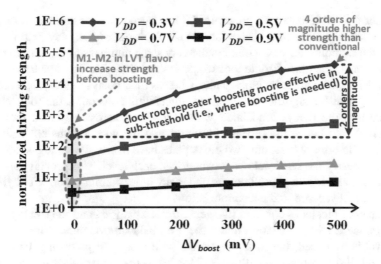

**Fig. 5.11** Driving strength of boostable clock repeater versus gate-boosted voltage at different $V_{DD}$ (normalized to conventional clock repeater with same transistor size)

## 5.5 Automated Design Flows for Reconfigurable Clock Networks and Integration with DVFS

### 5.5.1 Automated Clock Tree Design and Level Balance Principle

The insertion of bypassable and boostable repeaters can be fully automated by using commercial EDA tools, while making minor changes in the design flow and leaving the design effort nearly unaffected. Before the implementation of the reconfigurable clock network, bypassable and boostable repeaters need to be made available as standard cells. Hence, the existing standard cell library needs to be first enhanced with these few cells, which need to be laid out, characterized as cells (e.g., area, power, timing), and then imported into an existing digital design flow.

During clock tree synthesis (CTS), bypassable repeaters are used in place of conventional clock repeaters. CTS is run to automatically design the deepest clock network configuration at nominal voltage, which is carried out normally as if it were a conventional fixed network with non-bypassable repeaters. At nominal voltage, the overall delays of clock paths are inherently balanced by the place&route EDA tool, which optimizes the clock network to meet the targeted clock skew timing constraint. When $V_{DD}$ is reduced, repeater levels are bypassed since their wire delay becomes irrelevant, thus merging the previous and the subsequent wire into a single level. In general, repeater bypassing can upset the initially balanced clock paths delays, thus potentially eroding the clock skew reduction enabled by clock network reconfiguration. This potential issue is solved by introducing the concept of "level balance," as discussed below.

Under the "level balance" principle, CTS is purposely constrained to maintain the very same number of repeaters across all clock paths[2] (i.e., any path from the root to the endpoint leaves encounters the very same number of repeaters). Without this constraint, the place&route tool would instead limit itself to equalize only the overall delay of the different clock paths, whereas their delays in individual repeater levels might differ significantly. Instead, under this constraint, the tool tends to equalize also the delays associated with every level across all clock paths (see Fig. 5.12a), and hence the repeater strengths and the wire lengths are within the same level. In other words, any two clock paths have nearly the same overall delay, as well as nearly the same delay contribution at each level. This property allows to equalize the clock path overall delays under any other configuration with bypassed repeaters, as discussed in the example below.

As an example, let us consider two generic clock paths #1 and #2 as in Fig. 5.12b, based on the CTS optimization under the level balance principle. When a generic $i$-th level is bypassed due to its negligible wire delay, all preceding $1...(i-2)$-th levels and all subsequent $(i+1)$-th$...(N-2)$-th levels maintain exactly the same delay as in the original non-bypassed configuration. At the same time, the wire delay in the merged levels $(i-1)$ and $i$ is negligible, as this has indeed driven the merger of the two wires via the $i$-th level repeater bypassing. Accordingly, the only timing difference in paths #1 and #2 resulting from $i$th level repeater bypassing is due to the potentially different load seen by the repeater at the $(i-1)$-th level driving the two merged wires (highlighted in green in Fig. 5.12b). However, CTS under the level balance principle tends to maintain the strength of the repeaters at the $(i-1)$-th level nearly the same across clock paths, and the same applies to the length of the subsequent wires (and hence of their capacitive load). Accordingly, the delay of the merged $(i-1)$-th and $i$th levels remains nearly the same for the two clock paths, after repeater bypassing. This argument is immediately generalized to any pair of clock paths and across different combinations of bypassed repeaters. Accordingly, CTS under the level balance constraint helps minimize the clock path delay differences at each level, and hence minimize the overall clock path delay differences in any configuration.

It is worth noting that the above argument needs to be applied to any logic gate inserted in the clock network, not only repeaters. Accordingly, if some clock path contains cells that cannot be bypassed such as clock gaters (see Fig. 5.12c), dummy cells of the same type need to be added in all other branches in the same level to preserve the level balance principle (i.e., any clock path needs to encounter the very same number of cells, including gaters and other cells). This can be done either before or after CTS. Automated CTS is then performed to achieve the targeted clock skew and slope at nominal voltage, with reconfigurable repeaters being configured in normal mode.

---

[2] This can be done in either manual or automatic CTS mode in commercial tools. In manual CTS mode, this is achieved by explicitly specifying the number and the type of clock repeaters at each level. In auto CTS mode, the level-balanced option needs to be enabled in the clock specification file (e.g., add "LevelBalanced YES" in the .ctstch file in Cadence PnR tools).

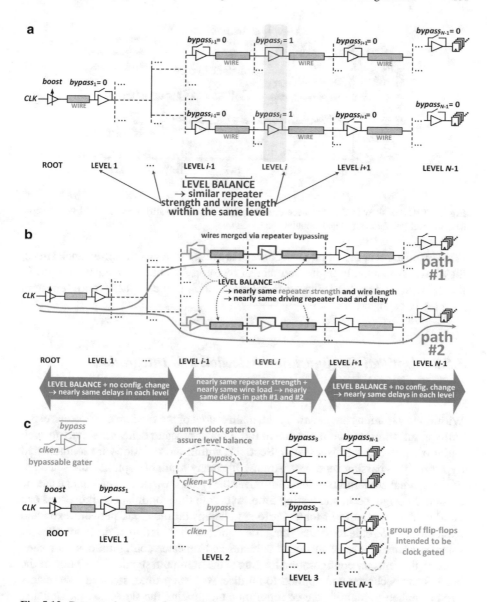

**Fig. 5.12** Dummy clock gater for level balance under clock gating

As further benefit, level-balanced clock trees are less sensitive to process and voltage corners, thanks to the equal number of repeaters (or other gates) in all clock paths. As an example, Fig. 5.13 shows the clock skew of a level-balanced and a level-imbalanced clock tree in the FFT accelerator in Sect. 5.6. Both clock trees were synthesized with the same constraints, with no level balance constraint in the imbalanced tree, which turned out to lead to a maximum imbalance of one repeater at any pair of clock paths. From Fig. 5.13, the level-balanced tree reduces the clock

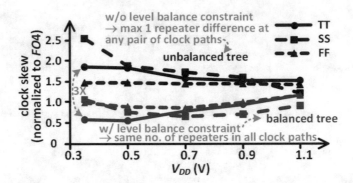

**Fig. 5.13** Clock skew in the FFT accelerator in Sect. 5.6 under various process and voltage corners in level-balanced and level-imbalanced clock tree

skew by up to 3× at the worst-case corner. As a disadvantage of level-balanced trees, they have a moderately higher number of repeaters,[3] and hence area and power. In the example in Fig. 5.13, the level-balanced clock tree uses 12% more repeaters than the imbalanced one.

### 5.5.2 Optimal Configuration Selection and Integration with DVFS

Within-die variations are generally the main source of the random clock skew, especially at sub-threshold voltages due to the stronger repeater delay sensitivity to such variations [30–32]. As discussed in Sect. 5.2, within-die variations are counteracted by properly bypassing repeaters. Among the many available options, the selection of the optimal configuration that minimizes the clock skew at a given voltage can be chosen at design, boot, and testing time. At design time, optimal configurations can be identified by running Monte Carlo simulations on the clock network extracted from the place&route tool, analyzing the statistical distribution of the clock skew in selected critical paths at different voltages and repeater configurations. For each voltage, the repeater configuration leading to the minimum standard deviation of the clock skew is chosen and applied to all dice. At testing time, selected test vectors can be applied to identify the configuration minimizing the skew, i.e., maximizing the clock frequency, discarding any configuration experiencing hold failures at each specific voltage. Similar considerations can be repeated at chip boot time.

Once the optimal configurations are identified for each operating supply voltage, they need to be dynamically assigned at run time based on the targeted voltage and frequency. Reconfigurable clock networks can be easily integrated into existing

---

[3] The same consideration holds for any other logic gate in the clock path, such as clock gaters. In this case, the presence of clock gaters in selected clock paths was balanced in other paths by adding a dummy clock gater in the same clock network level, to preserve the same number of logic gates across all clock paths.

dynamic voltage–frequency scaling (DVFS) schemes. Indeed, clock network reconfiguration at each voltage represents an immediate extension of the traditional look-up table that specifies the clock frequency at each $V_{DD}$. Figure 5.14 shows the detailed implementation of simultaneous voltage, frequency, and clock network reconfiguration adjustment at run time. The additional clock configuration column in the voltage–frequency look-up table sets the control signals required by the repeaters in Fig. 5.5, i.e. *bypass* and *boost*. Such additional clock network configuration column stores the optimal configurations identified at design, boot, or testing time, similarly to the clock frequency column. Lower voltages are associated with lower clock frequencies and shallower clock network configurations, as discussed in Sect. 5.2.

When voltage is dynamically scaled, clock frequency and repeater reconfiguration determine a transient that might lead to timing failures and hence malfunctioning, before the clock signal settles to its targeted and steady-state frequency. Several techniques are available to manage such transient in DVFS schemes [6, 28]. The number of pre-decoded configuration bits in the look-up table in Fig. 5.14 is equal to the number of clock repeater levels (i.e., several units or slightly more). Reconfiguration allows the granularity of a single repeater level. Each configuration bit is distributed to the repeaters (or gaters) lying in the corresponding level.

From the technology scaling perspective, the techniques discussed in this section become more advantageous when clock repeater variations become more pronounced. Hence, more substantial benefits are expectable in CMOS technologies with finer feature size.

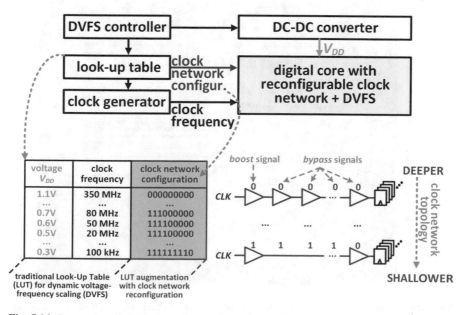

**Fig. 5.14** Integration of clock network reconfiguration in DVFS schemes, augmenting the voltage–frequency look-up table with clock configuration

## 5.6  Case Study: Reconfigurable Clock Network in FFT Accelerator

### 5.6.1  Testchip Design

To demonstrate the reconfigurable clock network concept, a 256-point 16-bit radix-4 complex Fast Fourier Transform processor based on the modified multipath delay commutator MDC architecture in [22, 23] was implemented in a 40 nm test chip. The die micrograph is shown in Fig. 5.15. The basic clock network design parameters are summarized in Table 5.1. The FFT processor was implemented in two versions: the first features the reconfigurable clock network, whereas the second was designed with a conventional clock network, as a baseline to experimentally quantify the benefits of reconfiguration. For fair comparison, both versions were designed with the very same design flow, clock network structure (e.g., number of repeater levels, number of buffers in each level), and timing constraints. According to the level balance principle in Sect. 5.5.1, both clock networks contain eight repeater levels. For both designs, the resulting clock skew at the nominal voltage was less than 1.3FO4 across all global corners after place&route. The sign-off clock skew and slope for conventional and reconfigurable clock network at nominal voltage are very close to each other, as shown by the clock phase delay map in Fig. 5.16.

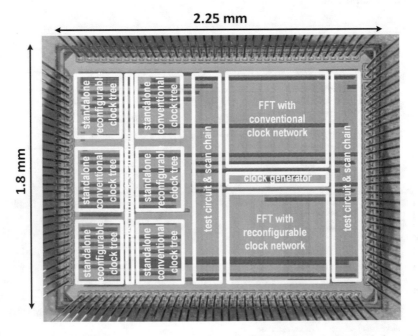

**Fig. 5.15** Chip micrograph and floor plan containing the FFT accelerators, the stand-alone clock networks, and the related testing harness

**Table 5.1** Parameters related to clock network design (FFT accelerator)

| | Clock network → | Conventional | Reconfigurable |
|---|---|---|---|
| Clock network | # Sinks | 20,283 | 20,283 |
| | # Repeater levels | 8 | 1–8 (tunable) |
| | # Repeaters | 1763 | 1763 |
| | Area of repeaters (%) | 1.4% | 3.2% |
| | Gate count | 10.4 k gates | 23.5 k gates |
| | Repeater cell area (24× strength) | 1× | 2.3× |
| General | Gate count | 729,000 | 742,000 |
| | Area of combinational/sequential/other cells | 61.8%/29.2%/9% | 60.9%/28.7%/10.4% |
| | Area overhead of reconfiguration (%) | – | 1.8% |
| | $V_{min}$ | 0.45 V | 0.34 V |
| | Min. energy | 2.49 nJ/FFT | 1.73 nJ/FFT |

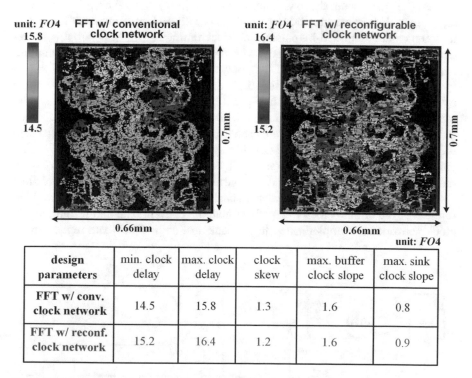

unit: *FO4* **FFT w/ conventional clock network**
15.8
14.5
0.7mm
0.66mm

unit: *FO4* **FFT w/ reconfigurable clock network**
16.4
15.2
0.7mm
0.66mm

unit: *FO4*

| design parameters | min. clock delay | max. clock delay | clock skew | max. buffer clock slope | max. sink clock slope |
|---|---|---|---|---|---|
| **FFT w/ conv. clock network** | 14.5 | 15.8 | 1.3 | 1.6 | 0.8 |
| **FFT w/ reconf. clock network** | 15.2 | 16.4 | 1.2 | 1.6 | 0.9 |

**Fig. 5.16** Clock phase delay map (top) and clock network parameters (bottom) of FFT designed with conventional (left) and reconfigurable clock network (right), from clock tree synthesis at nominal voltage (SS corner, worst-case RC extraction)

To enable further testing flexibility, the FFT accelerators in the test chip include an intentional skew injection mechanism to mimic different random variations to fairly compare different FFT cores at the same hold margin, in spite of inevitably different core-specific variations. As in Fig. 5.17, these paths end with an individual reconfigurable clock buffer as sink repeater, in which skew is injected by adjusting its delay through the tuning gate voltages $V_{tuning, p}$ and $V_{tuning, n}$. The delay of such few repeaters can be precisely controlled to cover the entire range of variations expected at all process corners and within-die variations. Skew injection was inserted in the most hold-critical paths, which were selected as the ones with (1) the lowest hold margin (i.e., lowest hold slack from post-P&R timing report) and (2) highest clock skew due to process variations (i.e., lowest number of clock buffers in common in the related clock paths). This path selection was performed through automated parsing of the post-P&R timing reports. Figure 5.18 shows the detailed procedure to properly select the hold-critical paths. First (Step a), the data paths in the post-place&route timing report are ranked by their hold slack, from the lowest to the highest (i.e., paths from the lowest hold margin to the highest). Then (Step b), starting from the lowest margin, the clock path pairs with the lowest number of repeaters in common are identified by parsing the clock timing report. In detail, this procedure allows to identify various paths with minimal hold margin and minimal number of clock repeaters in common (i.e., only one), as desired.

Once skew was injected as in Fig. 5.17, the resulting hold margin was measured by introducing the hold-fix buffer in Fig. 5.19, whose delay was tuned with a similar mechanism as in Fig. 5.17. Such tunable hold-fix buffers are inserted at hold-fix time by the place&route tool, and they are used in place of the hold-fix buffers available in the adopted commercial standard cell library. The hold time margin improvement gained with the skew reduction enabled by clock network reconfiguration is then measured as the hold-fix buffer delay change required to induce the first hold violation.

To gain a deep insight into the statistical distribution of the clock skew, 3456 replicas of the critical clock tree path of both the conventional and reconfigurable clock network were implemented in the same test chip. Clock path replicas were

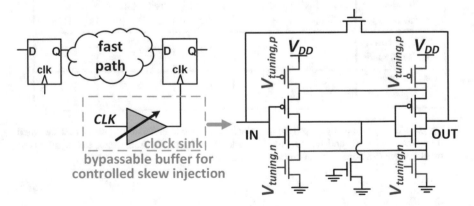

**Fig. 5.17** Bypassable repeater with skew injection in selected hold-critical paths

**Systematic procedure for critical path selection for skew injection**
**a) list hold-critical data paths with lowest hold margin  (rank from lowest to highest)**
**b) among these critical paths, find the paths with lowest number of clock repeaters in common (in the FFT example, skew is injected in various paths with minimal hold margin and only the first clock level in common)**

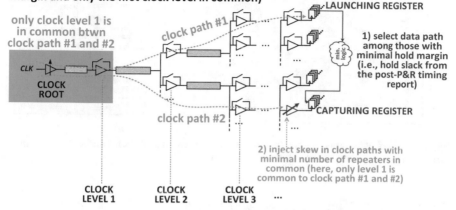

**Fig. 5.18**  Systematic procedure to select hold-critical paths for skew injection

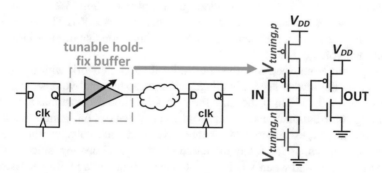

**Fig. 5.19**  Hold-fix buffer with tunable delay for hold margin tuning and measurement

extracted from the most hold-critical path in the FFT core (i.e., lowest hold margin), and implemented while maintaining the very same wire length and number of repeaters at every level. The clock path replicas are split into three banks as in Fig. 5.20. Each bank contains an entire eight-level clock tree with the same structure as the one in the FFT accelerators. The skew characterization of these clock path replicas was performed by embedding an on-chip time-to-digital converter. The latter was implemented by cascading a programmable delay line for coarse-grain measurements [33], and a Vernier time-to-digital converter for fine-grain measurements [34, 35]. Proper multiplexing is inserted to individually select each clock path replica and measure its skew with respect to other paths. To eliminate the delay mismatch caused by the multiplexing circuitry, a direct path from input to the fully balanced multiplexer tree was also created to characterize and subtract the multiplexer delay mismatch across different branches (see green line in Fig. 5.20).

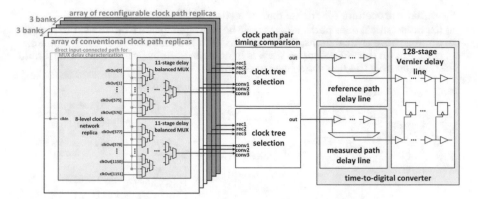

**Fig. 5.20** Clock path replicas of reconfigurable and conventional clock networks, and skew measurement circuitry based on time-to-digital conversion

## 5.6.2   Clock Skew Measurement Results

The measurement results from the clock path replicas in Fig. 5.20 are plotted in Fig. 5.21 in the form of clock skew standard deviation versus $V_{DD}$. This figure shows that there is always an optimal clock network configuration that minimizes the clock skew at each voltage.

As expected from Sect. 5.2, the conventional clock network with no bypassed level exhibits the lowest skew at voltages around the nominal voltage, and in particular for $V_{DD} > 0.7$ V. This is because the wire delay time constant makes an important timing contribution at above-threshold voltages, whose mitigation requires the insertion of the largest number of repeater levels. At such voltages, the boostable repeater strength cannot be really increased through gate boosting, since limited or no boosting is allowed when $V_{DD}$ is close to the nominal voltage, for transistor reliability reasons. In other words, the clock root repeater is conventionally sized to meet the clock skew and slope targets in the first level of the clock network, without the support of gate boosting.

At the lower end of the voltage range and hence the sub-threshold region (down to 0.3 V), the configuration with minimum skew has all intermediate levels being bypassed, and the root (sink) repeater configured in boost (normal) mode (named 111111110 in Fig. 5.5). When $V_{DD}$ increases above 0.3 V, the optimal configuration expectedly has fewer bypassed levels. For example, the optimal configuration for 0.5 V and 0.6 V is 111100000, for 0.7 V it is 111000000.

The histogram of the clock skew distribution between pairs of clock path replicas is depicted in Fig. 5.22a, for both the conventional and the reconfigurable clock network. The resulting clock skew reduction enabled by the reconfigurable clock network is plotted versus $V_{DD}$ in Fig. 5.22b. In this plot, the standard deviation of the clock skew in the conventional network is normalized to the corresponding value in the reconfigurable one.

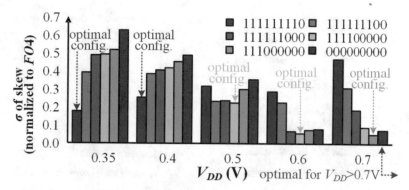

**Fig. 5.21** Measured clock skew normalized to FO4 vs. $V_{DD}$ for various clock network configurations. The optimal configuration with minimal clock skew is highlighted for each voltage

Figure 5.22b shows that the reconfigurable network with no bypassed repeater has very similar skew as the conventional one at nominal $V_{DD}$, since the reconfigurable repeater variability is close to the conventional one as was observed in Sect. 5.3. From Fig. 5.22b, repeater bypassing starts being beneficial at 0.7 V and below and expectedly offers the maximum clock skew reduction at the lowest voltage. Overall, the clock skew is reduced by up to 3.3×.

As a design example, the FFT accelerator described in Sect. 4.1 was designed, manufactured, and tested to evaluate the benefits of reconfigurable clock networks in terms of hold margin and minimum operating voltage $V_{min}$. The hold margin was measured by tuning the hold-fix buffer delay at the point of first hold failure, as determined by the specific variations in each testchip. To fairly compare the conventional and the reconfigurable clock network under optimal configuration, controlled skew was injected to set the first point of failure at the same voltage of $V_{DD} = 0.45$ V in both the conventional and the reconfigurable one with the equivalent 000000000 configuration.

The optimal configuration identified in the reconfigurable clock network is plotted versus $V_{DD}$ in Fig. 5.23. From this figure, the hold margin of the FFT accelerator is improved by up to 2.5 standard deviations of the clock skew at $V_{DD} < 0.5$ V, compared to the conventional clock network. The standard deviation of the clock skew was experimentally evaluated as discussed in Sect. 5.6.1. Comparing Figs. 5.22 and 5.23, the optimal configurations of the FFT accelerator are confirmed to be the same as the ones identified in the array of clock network replicas.

The voltage range in which each clock network configuration ensures correct operation with no timing violations (i.e., positive setup and hold margin) is plotted in Fig. 5.24. This range is expectedly shifted to lower voltages, when more levels are bypassed. The lower end $V_{min}$ of this range in the 000000000 configuration was set to 0.45 V via controlled skew injection, as discussed above. $V_{min}$ at configurations with more bypassed repeater levels is expectedly reduced, thanks to the skew reduction enabled by the clock network reconfiguration as in Fig. 5.23. Quantitatively, the reconfigurable clock network under configuration 111111110 reduces $V_{min}$ down

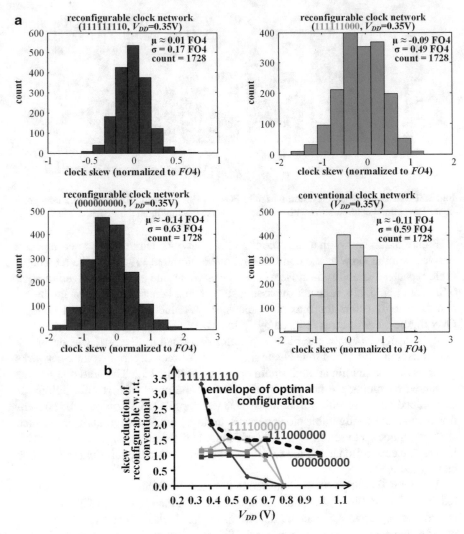

**Fig. 5.22** (**a**) Histogram of clock skew measured between all clock path replicas under various configurations and (**b**) clock skew reduction factor of reconfigurable clock network under optimal configuration w.r.t. conventional (i.e., $\sigma_{conventional}/\sigma_{reconfigurable}$) at 0.35 V

to 0.34 V, as compared to 0.45 V of the conventional one. As an expectable result, the maximum operating voltage $V_{max}$ of configurations is also progressively reduced. This is because the chosen clock root repeater size is not sufficient to drive the additional bypassed levels at larger $V_{DD}$, at which gate boosting is less effective. It is worth noting that this is actually not a limitation, as the clock root repeater size can be always increased to make its strength adequate for the most critical configuration

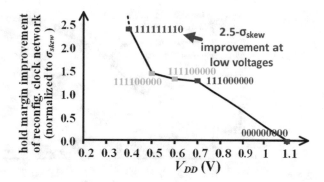

**Fig. 5.23** Hold margin improvement of reconfigurable clock network compared to conventional clock network (colors as in Figs. 5.21 and 5.22)

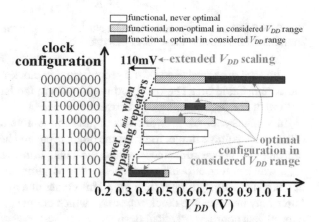

**Fig. 5.24** $V_{DD}$ range in which positive setup and hold margin are assured for each configuration, and voltage range in which the configuration minimizes the clock skew (colors as in Figs. 5.21 and 5.22)

at any voltage, down to the lowest end of the $V_{DD}$ range, as discussed in Sect. 5.4. In any case, $V_{max}$ is always well above the voltage range in which the configuration is optimal, as shown in Fig. 5.24. This further confirms that the voltage upper bound in each configuration is not an actual limit in practical cases where the configuration is selected to minimize the clock skew.

### 5.6.3 Improvements in Performance, Robustness, and Energy Offered by Reconfigurable Clock Networks

The clock skew reduction in the previous subsection directly translates into performance, robustness, and energy benefits, as discussed in Sect. 5.1. Thanks to the 110 mV reduction in $V_{min}$ shown in Fig. 5.24, the minimum energy of the FFT accelerator in Fig. 5.25 is further reduced by 31% when $V_{DD}$ is scaled down to 0.34 V, compared to the fixed deep configuration having $V_{min} = 0.45$ V. In other words, clock reconfiguration extends operation at lower supply voltages, and hence offers additional opportunities to reduce the energy. At the same time, the maximum frequency is increased by up to 8% in the near- and sub-threshold regions.

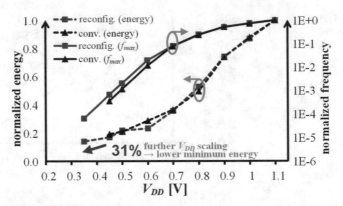

**Fig. 5.25** Measured energy and maximum clock frequency of the FFT accelerator vs. $V_{DD}$ under conventional and reconfigurable clock network

The total area overhead of the reconfigurable clock network is only 1.8% from the data in Table 5.1. The overhead is only due to the larger area of reconfigurable buffers in Fig. 5.6, compared to conventional ones.

In summary, clock network reconfiguration is able to mitigate the clock skew over a wide voltage range, improve the clock cycle and hence performance, the hold margin, and hence robustness. Also, the latter improvement significantly lowers $V_{min}$, which in turn allows further energy downscaling, or further yield improvement at a targeted voltage higher than $V_{min}$. This comes at a minor area overhead since the latter only involves the clock repeaters, which are generally a small fraction of the overall gate count in practical designs.

## 5.7  Conclusion

In this chapter, reconfigurable clock networks have been introduced to dynamically optimize the number of clock repeater levels to minimize the clock skew under a wide voltage range. Clock repeaters are bypassed to progressively reduce the number of clock levels and mitigate the skew degradation, when $V_{DD}$ is scaled down. The integration of the proposed approach within an automated digital design flows and run-time dynamic voltage–frequency scaling have been shown to be straightforward.

Based on the FFT accelerator test vehicle in Sect. 4.1, clock skew mitigation up to 3.3× has been demonstrated across a wide voltage range, without sacrificing skew at any specific $V_{DD}$. In contrast, conventional fixed clock networks designed at a given $V_{DD}$ need to sacrifice clock skew at voltages that are different from the supply adopted during the clock tree synthesis. Clock reconfiguration improves robustness against hold time violations and hence further reduces $V_{min}$, or equivalently it improves the yield at low $V_{DD}$. $V_{min}$ reduction of 110 mV and 31% minimum energy reduction have been experimentally shown. The area overhead has been shown to be minimal (1.8%).

As further benefit, clock reconfiguration also facilitates timing closure at design time, as the clock skew at different $V_{DD}$ settings no longer imposes conflicting requirements on the clock network.

# References

1. L. Lin, S. Jain, M. Alioto, Reconfigurable clock networks for random skew mitigation from subthreshold to nominal voltage, in *IEEE ISSCC Digest of Technical Papers, San Francisco (CA)*, (2017), pp. 440–441
2. L. Lin, S. Jain, M. Alioto, Reconfigurable clock networks for wide voltage scaling. IEEE J. Solid State Circuits **54**(9), 2622–2631 (2019)
3. M. Alioto (ed.), *Enabling the Internet of Things—From Integrated Circuits to Integrated Systems* (Springer, Berlin, 2017)
4. T. Xanthopoulos (Ed.), *Clocking in Modern VLSI Systems*, 2009.
5. M. Alioto, E. Consoli, G. Palumbo, *Flip-Flop Design in Nanometer CMOS—From High Speed to Low Energy* (Springer, Berlin, 2015)
6. T. Burd, T. Pering, A. Stratakos, R. Brodersen, A dynamic voltage scaled microprocessor system, in *IEEE ISSCC Digest of Technical Papers*, (2000)
7. S. Jain, S. Khare, S. Yada, V. Ambili, P. Salihundam, S. Ramani, S. Muthukumar, M. Srinivasan, A. Kumar, S. Kumar, R. Ramanarayanan, V. Erraguntla, J. Howard, S. Vangal, S. Dighe, G. Ruhl, P. Aseron, H. Wilson, N. Borkar, V. De, S. Borkar, A 280 mV-to-1.2 V wide-Operating-range IA-32 processor in 32 nm CMOS, in *IEEE ISSCC Digest of Technical Papers, San Francisco (CA)*, (2012)
8. W. Wang, P. Mishra, System-wide leakage-aware energy minimization using dynamic voltage scaling and cache reconfiguration in multitasking systems. IEEE Trans. VLSI Syst. **20**(5) (2012)
9. A.P. Chandrakasan, D.C. Daly, D.F. Finchelstein, J. Kwong, Y.K. Ramadass, M.E. Sinangil, V. Sze, N. Verma, Technologies for ultradynamic voltage scaling. Proc. IEEE **98**(2), 191–214 (2010)
10. M. Seok, D. Jeon, C. Chakrabati, D. Blaauw, D. Sylvester, Extending energy-saving voltage scaling in ultra low voltage integrated circuit designs, in *Proceedings of ICICDT, Austin (TX)*, (2012)
11. D. Jacquet, F. Hasbani, P. Flatresse, R. Wilson, F. Arnaud, G. Cesana, T.D. Gilio, C. Lecocq, T. Roy, A. Chhabra, C. Grover, O. Minez, J. Uginet, G. Durieu, C. Adobati, D. Casalotto, F. Nyer, P. Menut, A. Cathelin, I. Vongsavady, P. Magarshack, A 3 GHz dual core processor ARM Cortex TM-A9 in 28 nm UTBB FD-SOI CMOS with ultra-wide voltage range and energy efficiency optimization. IEEE J. Solid State Circuits **49**(4), 812–826 (2014)
12. F. Abouzeid, S. Clerc, B. Pelloux-Prayer, F. Argoud, P. Roche, 28 nm CMOS, energy efficient and variability tolerant, 350 mV-to-1.0 V, 10 MHz/700 MHz, 252 bits Frame Error-Decoder, in *Proceedings of ESSCIRC*, (2012), pp. 153–156
13. S. Hsu, A. Agarwal, M. Anders, S. Mathew, H. Kaul, F. Sheikh, R. Krishnamurthy, A 280 mV-to-1.1V 256b reconfigurable SIMD vector permutation engine with 2-dimensional shuffle in 22 nm CMOS, in *IEEE ISSCC Digest of Technical Papers, San Francisco (CA)*, (2012)
14. M. Seok, D. Blaauw, D. Sylvester, Robust clock network design methodology for ultra-low voltage operations. IEEE J. Emerg. Select. Topics Circuits Syst. **1**(2), 120–130 (2011)
15. J.R. Tolbert, X. Zhao, S.K. Lim, S. Mukhopadhyay, Analysis and design of energy and slew aware subthreshold clock systems. IEEE Trans. Comput. Aided Design Integr. Circuits Syst. **30**(9), 1349–1358 (2011)

16. X. Zhao, J.R. Tolbert, S. Mukhopadhyay, S.K. Lim, Variation-aware clock network design methodology for ultralow voltage (ULV) Circuits. IEEE Trans. Comput. Aided Design Integr. Circuits Syst. **31**(8), 1222–1234 (2012)
17. C. Sitik, W. Liu, B. Taskin, E. Salman, Design methodology for voltage-scaled clock distribution networks. IEEE Trans. VLSI Syst. **24**(10), 3080–3093 (2016)
18. S. Kim, M. Seok, Reconfigurable regenerator-based interconnect design for ultra-dynamic-voltage-scaling systems, in *Proceedings of ISLPED 2014, La Jolla (CA)*, (2014), pp. 99–104
19. J. Wang, N. Pinckney, D. Blaauw, D. Sylvester, Reconfigurable self-timed regenerators for wide-range voltage scaled interconnect, in *2015 IEEE Asian Solid-State Circuits Conference (A-SSCC), Xiamen (China)*, (2015), pp. 1–4
20. M. Alioto, G. Scotti, A. Trifiletti, A novel framework to estimate the path delay variability on the back of an envelope via the Fan-Out-of-4 Metric. IEEE Trans. CAS Pt. I **64**(8), 2073–2085 (2017)
21. C. Augustine, C. Tokunaga, A. Malavasi, A. Raychowdhury, M. Khellah, J. Tschanz, V. De, Characterization of PVT variation & aging induced hold time margins of flip-flop arrays at NTV in 22nm tri-gate CMOS, in *Proceedings of IEDM*, (2016), pp. 894–897
22. S. Jain, L. Lin, M. Alioto, Dynamically adaptable pipeline for energy-efficient microarchitectures under wide voltage scaling. IEEE J. Solid State Circuits **53**(2), 632–641 (2018)
23. D. Jeon, M. Seok, C. Chakrabarti, D. Blaauw, D. Sylvester, A super-pipelined energy efficient subthreshold 240 MS/s FFT core in 65 nm CMOS. IEEE J. Solid State Circuits **47**(1) (2012)
24. S. Hanson, B. Zhai, K. Bernstein, D. Blaauw, A. Bryant, L. Chang, K.K. Das, W. Haensch, E.J. Nowak, D.M. Sylvester, Ultralow-voltage, minimum-energy CMOS. IBM J. Res. Dev. **50**(4/5) (2006)
25. D. Bol, J.D. Vos, C. Hocquet, F. Botman, F. Durvaux, S. Boyd, D. Flandre, J. Legat, SleepWalker: A 25-MHz 0.4-V Sub-mm$^2$ 7uW/MHz Microcontroller in 65-nm LP/GP CMOS for low-carbon wireless sensor nodes. IEEE J. Solid State Circuits **48**(1), 20–32 (2013)
26. J. Myers, A. Savanth, R. Gaddh, D. Howard, P. Prabhat, D. Flynn, A subthreshold ARM Cortex-M0+ Subsystem in 65 nm CMOS for WSN applications with 14 power domains, 10T SRAM, and integrated voltage regulator. IEEE J. Solid State Circuits **51**(1), 31–44 (2016)
27. C. Tokunaga, J. F. Ryan, C. Augustine, J. P. Kulkarni, Y. Shih, S. T. Kim, R. Jain, K. Bowman, A. Raychowdhury, M. M. Khellah, J. W. Tschanz, V. De, "A Graphics Execution Core in 22nm CMOS featuring adaptive clocking, selective boosting and state-retentive sleep," in *IEEE ISSCC Digest of Technical Papers*, San Francisco (CA), 2014, pp. 108-109.
28. M. Keating, D. Flynn, A. Gibbons, R. Aitken, K. Shi, *Low Power Methodology Manual For System-on-Chip Design* (Springer, Berlin, 2007)
29. M. Alioto, Ultra-low power VLSI circuit design demystified and explained: a tutorial. IEEE Trans. Circuits Syst. Pt. I **59**(1), 3–29 (2012)
30. M. Alioto, G. Palumbo, M. Pennisi, Understanding the effect of process variations on the delay of static and domino logic. IEEE Trans. VLSI Syst. **18**(5), 697–710 (2010)
31. M. Eisele, J. Berthold, D. Schmitt-Landsiedel, R. Mahnkopf, The impact of intra-die device parameter variations on path delays and on the design for yield of low voltage digital circuits. IEEE Trans. VLSI Syst. **5**(4), 360–368 (1997)
32. K. Bowman, S. Duvall, J. Meindl, Impact of Die-to-Die and within-die parameter fluctuations on the maximum clock frequency distribution for gigascale integration. IEEE J. Solid State Circuits **37**(2), 183–190 (2002)
33. N. Nedovic, W.W. Walker, V.G. Oklobdzija, A test circuit for measurement of clocked storage element characteristics. IEEE J. Solid State Circuits **39**(8), 1294–1304 (2004)
34. T.E. Rahkonen, J.T. Kostamovaara, The use of stabilized CMOS delay lines for the digitization of short time intervals. IEEE J. Solid State Circuits **28**(8), 887–894 (1993)
35. P. Dudek, S. Szczepanski, J.V. Hatfield, A high-resolution CMOS time-to-digital converter utilizing a Vernier delay line. IEEE J. Solid State Circuits **35**(2), 240–247 (2000)

# Chapter 6
# Conclusions

**Abstract**  In the past two decades, the coverage of a progressively wider range of power-performance design points has led to the introduction of wide voltage scaling, where the supply voltage is dynamically adjusted from nominal down to near- or sub-threshold voltages. This has allowed to achieve low power in the common case of low-performance targets, while still being able to meet a higher peak performance when occasionally required. However, simultaneous minimization for high- and low-performance targets has proved to be inherently challenging. Indeed, improving power on one side of the voltage range invariably degrades power and performance at the other end, in conventional designs with fixed microarchitecture and circuit implementation.

**Keywords**  Wide voltage scaling · Near-threshold · Sub-threshold · Microarchitecture · Reconfiguration · Pipestage · Data path · Clock path · Re-pipelining · Application specific hardware · Logic depth · Thread-level reconfiguration · Microprocessors · Row aggregation · SRAM array · EDA tool · Decoder · Post-layout simulations · Clock tree topology · Clock network · Statistical skew · Fixed clock networks · Joint repipelining · Register bypassing · Accelerators · Time interleaving · Parallelism · Gate-level netlist · Skew · Design flow · Dynamic voltage frequency scaling · Minimum energy · Peak performance · Minimum energy point (MEP) · CAD algorithm

In the past two decades, the coverage of a progressively wider range of power-performance design points has led to the introduction of wide voltage scaling, where the supply voltage is dynamically adjusted from nominal down to near- or sub-threshold voltages. This has allowed to achieve low power in the common case of low-performance targets, while still being able to meet a higher peak performance when occasionally required. However, simultaneous minimization for high- and low-performance targets has proved to be inherently challenging. Indeed, improving power on one side of the voltage range invariably degrades power and performance at the other end, in conventional designs with fixed microarchitecture and circuit implementation.

This book introduces novel digital circuit techniques and design methodologies to enable fine-grain reconfiguration down to the pipestage level in the data path, and the repeater level in the clock path. Such techniques and methodologies further extend

© Springer Nature Switzerland AG 2020
S. Jain et al., *Adaptive Digital Circuits for Power-Performance Range beyond Wide Voltage Scaling*, https://doi.org/10.1007/978-3-030-38796-9_6

the power, energy, and performance range beyond allowed by wide voltage scaling. Synergistic data/clock path reconfiguration and voltage scaling permits to meet a peak performance that is higher than pure voltage scaling at nominal voltage, as well as a minimum energy lower than the minimum energy point under low voltage.

Data and clock path reconfiguration is shown to make digital designs highly versatile and energy efficient under a wide range of design targets, to reduce the consumption in the common case, while making the digital sub-system faster and more responsive under event occurrence. At the same time, the wider power-performance tradeoff enables more extensive reuse of a digital design instance across a wide range of applications and systems on chip. This circumvents the traditional designer's dilemma of choosing which end of the power-performance spectrum should be favored over the other, when adopting wide voltage scaling. Drop-in solutions for fully automated and low-effort design based on commercial design tools are extensively discussed for processors, accelerators, and on-chip memories. Ultimately, the extension of the power-performance tradeoff through data/clock reconfiguration keeps the design effort nearly unaltered, making its integration into existing design flow seamless.

In Chap. 2, three techniques were introduced to enable microarchitecture reconfiguration in the data path, further extending the power-performance benefits of conventional wide voltage scaling. First, pipeline-level reconfiguration was introduced for designs where re-pipelining is straightforward, as in the case of application-specific hardware. This technique serves as a very effective knob to achieve variable logic depth to cater to its conflicting requirement for energy-optimal operation over a wide range of supply voltages. As a second technique, thread-level reconfiguration was introduced for general-purpose architectures (e.g., microprocessors) where the critical path may lie in a feedback loop, and re-pipelining would require major redesign of the control flow and the software stack and introduce new data dependencies and hence hazards that would degrade both power and performance. In such cases, time interleaving has been selectively introduced to achieve variable logic depth. In both pipestage- and thread-level reconfiguration, the overhead is expectedly low since they both modify the registers only, while keeping the combinational logic the same. As third technique, the associated on-chip SRAM has been made reconfigurable to further increase its performance beyond the nominal voltage, to support the performance boost of the logic circuitry obtained via either pipestage- or thread-level reconfiguration. SRAM reconfiguration has been achieved through selective row aggregation.

In Chap. 3, fully automated data path design methodologies have been introduced for standard cell-based designs based on pipestage- and thread-level reconfiguration, as well as for SRAM array designs. As common goal, drop-in solutions for existing architectures that allow the above capability at very low design effort have been described. Regarding the design methodologies for logic, they are based on gate-level manipulations and are hence architecture-agnostic and can be adopted regardless of whether the design is described at the RTL level, or it is externally provided by a soft IP vendor. Such methodologies rely on commercial EDA tools and scripts that implement graph algorithms to ultimately integrate such tools into a unitary design flow. Regarding SRAM memories, low-effort design methodologies to selectively boost up the performance beyond allowed by nominal voltage have been introduced. The selective row aggregation technique has been shown to allow full reuse of the vast majority of existing SRAM arrays generated by memory compilers, as only a minor change in the decoder is needed (i.e., a single gate).

In Chap. 4, data path pipestage-, thread-, and memory bank-level reconfigurations have been explored through several test vehicles such as FFT accelerator, FIR filter, fixed point multiplier, and ARM Cortex-M0 system including memory, through testchip characterization, and post-layout simulations. Pipeline-level reconfiguration has been shown to yield an energy and throughput benefit of up to 38 and 80% at maximum area overhead of 9%. Thread-level reconfiguration in processors has been shown to yield a maximum throughput improvement by 80% at nominal voltage, at slightly area overhead of 16.4%. The stand-alone memory access time improvement was measured to be up to 85%. The overall system minimum energy has been reduced by up to 40%.

In Chap. 5, clock path reconfiguration has been discussed to keep random clock skew under control over a wide range of supply voltages. Especially under a wide voltage range, the clock network indeed significantly affects the timing/power closure, the range of voltages assuring correct operation and in particular $V_{min}$, and the performance. At the same time, wide voltage scaling imposes conflicting requirement on the clock tree topology selection at the two ends of the voltage range. A conventional fixed clock network designed at a given voltage leads to clock skew degradation at the other end of the voltage range, and hence to a degradation in robustness, performance, and energy efficiency. Accordingly, clock network reconfiguration is introduced to minimize the clock skew over a wide voltage range, dynamically adapting the number of repeater levels to the operating voltage to implement the clock network structure that best fits the supply voltage. This approach has been verified on a testchip with an FFT accelerator and a stand-alone clock tree for statistical skew characterization. The measurement results have shown that reconfigurable clock networks can reduce clock skew by up to 3.3×, enabling more than 100 mV $V_{min}$ reduction and 31% minimum energy reduction at 1.8% area penalty, compared to traditional fixed clock networks.

Across the previous chapters and the underlying publications from the same authors [1–7] (others are under way), the following fundamental insights have been gained:

- The conflicting logic depth requirements for energy-optimal microarchitectures over a wide voltage range can be reconciled via bypassable registers. This is obtained via either re-pipelining or time interleaving and selective replacement of proper conventional registers. Register reconfiguration can be easily integrated in run-time power management schemes such as DVFS.
- Joint re-pipelining and register bypassing enables pipeline-level reconfiguration, which is appropriate for microarchitectures that can be crafted to avoid data dependencies (e.g., accelerators). Joint time interleaving and register bypassing enables thread-level reconfiguration, which is more appropriate for general-purpose microarchitectures with legacy constraints (e.g., microprocessors), where automated re-pipelining is difficult. This technique is not necessarily restricted to microprocessors and can be generalized to designs with critical paths in feedback loops and cannot be re-pipelined easily. Thread-level

reconfiguration is functionally equivalent to scalable level of parallelism (or hardware replication).

- Microarchitecture reconfiguration is achieved through fully automated design flows based on commercially available EDA tools, and scripts that have been made publicly available by the authors (see Appendix).
- Microarchitecture reconfiguration is directly applied to gate-level netlists and is immediately extended to local standard cell memories based on latches or flip-flops, to ultimately achieve consistent speed-up in both memory and logic. The more commonly adopted SRAM memories require the introduction of a different reconfiguration mechanism based on row aggregation. The implementation is straightforward and requires only the replacement of a logic gate in the row decoder of each bank.
- The conflicting wire and delay requirements for clock networks with nearly minimal skew over a wide voltage range can be reconciled via bypassable repeaters. The resulting network topology reconfiguration can be fully supported by commercial tools through minor design flow changes and can be straightforwardly integrated with conventional DVFS. The same concepts are immediately generalized to the distribution of other global signals, not being restricted to the clock signal.
- Further reduction in the minimum energy is enabled by $V_{min}$ reduction through reconfigurable clock networks and leakage/dynamic energy optimization through simultaneous microarchitecture reconfiguration.

With this book and the help of the scripts in the Appendix, we really hope that our work will be useful to the reader to further expand the horizons of microarchitecture reconfiguration and further advance the state of the art in both energy efficiency and peak performance. Hopefully, this will ultimately enable new applications that our community and society can truly benefit from.

# References

1. S. Jain, L. Lin, M. Alioto, Dynamically adaptable pipeline for energy-efficient microarchitectures under wide voltage scaling. IEEE J. Solid State Circuits **53**(2), 632–641 (2018)
2. S. Jain, L. Lin, M. Alioto, Drop-in energy-performance range extension in microcontrollers beyond VDD scaling, in *IEEE Asian Solid-State Circuits Conference (A-SSCC)*, (2019)
3. L. Lin, S. Jain, M. Alioto, Reconfigurable clock networks for random skew mitigation from subthreshold to nominal voltage, in *IEEE ISSCC Dig. Tech. Papers, San Francisco (CA)*, (2017), pp. 440–441
4. L. Lin, S. Jain, M. Alioto, Reconfigurable clock networks for wide voltage scaling. IEEE J. Solid State Circuits **54**(9), 2622–2631 (2019)
5. S. Jain, L. Lin, M. Alioto, Design-oriented energy models for wide voltage scaling down to the minimum energy point. IEEE Trans. Circuits Syst. Pt. I **64**(12), 3115–3125 (2017)
6. S. Jain, L. Lin, M. Alioto, Automated design of reconfigurable micro-architectures for accelerators under wide voltage scaling. In print on *IEEE Transactions of Very Large Scale Integrated Circuits*
7. M. Alioto (ed.), *Enabling the Internet of Things—From Integrated Circuits to Integrated Systems* (Springer, Berlin, 2017)

# Appendix

This section provides a detailed description of how to build a complete automated design flow based on the techniques discussed in Chaps. 1–6, utilizing only commercially available EDA tools and the dedicated scripts discussed below. The resulting flow is architecture-agnostic and allows to reconfigure logic at pipeline and thread level. The dedicated scripts have been written in Python and C++. The Python scripts mainly handle text-level parsing of the netlist and implement lightweight netlist manipulation algorithms. The C++ scripts instead handle CPU-intensive and complex graph algorithms for more sophisticated netlist manipulations.

The complete flow and usage of scripts for pipeline- and thread-level reconfiguration is shown in Fig. A.1. The green color in the flow is associated with scripts (or output of the scripts). The black color refers to synthesis (e.g., Design Compiler) or place&route tool in the flow (e.g., Innovus). The flow for pipeline-level reconfiguration is described in Sect. A.1, whereas the thread-level reconfiguration is detailed in Sect. A.2. Instructions are provided in Sect. A.3 on how the scripts are organized and can be retrieved from our public repository.

## A.1 Pipeline-Level Reconfiguration

As a preliminary step, the available behavioral RTL is first synthesized to generate the post-synthesis gate-level netlist. As an alternative design scenario, such netlist might also be available in the form of a soft IP from a third-party vendor.

The post-synthesis netlist then regularly goes through the flow described in Chap. 3 and in particular in Fig. 3.1a. At Step 1 in this figure, the design is re-pipelined if the existing design does not meet the maximum throughput requirement at the nominal voltage. It is instead left unaltered, if the requirement is already met by the original netlist. Re-pipelining is carried out through the *repipeline.py* script in Fig. A.1, which manipulates the gate-level netlist to insert a specified number of

© Springer Nature Switzerland AG 2020
S. Jain et al., *Adaptive Digital Circuits for Power-Performance Range beyond Wide Voltage Scaling*, https://doi.org/10.1007/978-3-030-38796-9

# PIPELINE-LEVEL RECONFIGURATION FLOW

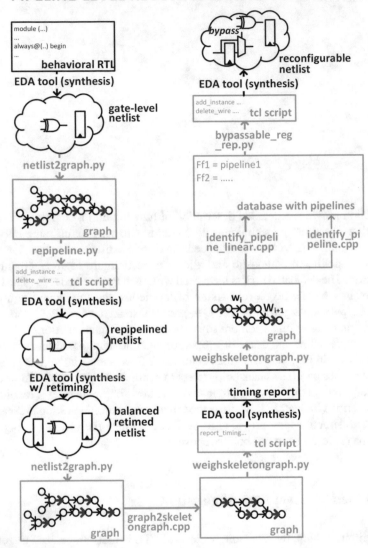

**Fig. A.1** Reconfiguration flow detailing the usage of dedicated scripts for pipeline-level reconfiguration

registers from the input through a Tcl script, which is inherently compatible with any well-established commercial synthesis tool [1, 2].

The re-pipelined netlist is then retimed using the exiting retiming features in commercial synthesis tools (Step 2 in Fig. 3.1a), as shown in Fig. A.1. The final retimed netlist is passed to the *netlist2graph.py* script that converts it to a *netlistgraph* (see Step 3.1.1 in Fig. 3.5), as summarized in Fig. A.1. The resulting *netlistgraph* is then passed to the *graph2skeletongraph.cpp* script, which converts *netlistgraph* into *skeletongraph* (Step 3.1.2 in Fig. 3.5) [2].

The skeleton graph is weighted through the *weighskeletongraph.py* script in Fig. A.1, which corresponds to Step 3.2 in Fig. 3.1a. This script generates the Tcl script that invokes the place&route tool to let it generate the timing report. The same *weighskeletongraph.py* script also parses the timing report and back annotates the weights to the skeleton graph [2].

The weighted skeleton graph is then used by the *identify_pipeline.cpp* (*identify_pipeline_linear.cpp*) script in the general case of non-linear pipelines (simpler case of linear pipelines). The script identifies the bypassable registers, corresponding to Steps 3.3 and 3.4 in Fig. 3.1a [2]. The resulting output database of bypassable registers is then used by the *bypassable_reg_rep.py* script to replace the corresponding conventional registers by bypassable ones. In particular, this script generates a Tcl script that is then run by the synthesis or place&route tool to actually carry out the replacement. The bypassable registers are also clock gated via same script [1, 2], as explained in Sect. 3.8.

## A.2    Thread-Level Reconfiguration

Thread-level reconfiguration follows the similar flow in Fig. 3.1b, whose detailed scripts and tools are described in Fig. A.2. The available gate-level netlist is passed to the *repipeline.py* script, which was used above for re-pipelining by setting an internal flag. When the flag is set to time interleaving mode, the same script generates a Tcl script that is successively run by the synthesis (or place&route) tool to replicate all the registers by the specified number of input channels $N$ to be time interleaved [3], as defined in Sect. 2.6 (e.g., 2 in the examples in Sects. 2.7 and 4.3). This implements Step 1 in Fig. 3.1b.

The resulting time-multiplexed netlist is then retimed using the widely available retiming capabilities of commercial synthesis tools (Step 2 in Fig. 3.1b), as shown specifically in Fig. A.2 [3]. This step approximately splits each pipestage in the original design into a number of pipestages $N$, as specified above in the *repipeline. py* script.

The resulting retimed netlist is passed to the *twoslow2oneslow.py* script, which performs the same function as the *identify_pipeline.cpp* script for non-linear pipelines (*identify_pipeline_linear.cpp* for linear pipelines). In detail, this script identifies the set of registers that should be made bypassable, based on the criteria in Sect. 3.9 and Fig. 3.1b (i.e., bypass all replica registers introduced in the baseline, so that their bypassing bring the reconfigurable microarchitecture back to the baseline one [3]).

The resulting database is used by the *bypassable_reg_rep.py* script to replace all the identified registers by their bypassable counterpart [3] (Step 4 in Fig. 3.1b). This script again generates a Tcl scripts compatible with commercial synthesis (or place&route) tool to actually carry out this replacement. The bypassable registers are also clock gated via same script.

## A.3   Detailed Description of Script I/Os

This section describes all the inputs and outputs associated with each script, as well as syntax instructions to run it. In this open-source set of scripts [4], the sample netlist and the designs have been created using the open-source standard cell libraries available at [5]. The adoption of this standard cell library overcomes the limitations that would otherwise be imposed by non-disclosure agreements associated with commercial process design kit (PDK). The set of scripts can be easily ported to any commercially available PDK by changing the environment variables library files, technology file names, and the name of standard cells for flip-flops, clock gaters, and multiplexers.

1. **repipeline.py:** The script is used to perform re-pipelining [6] or introduce time interleaving [3] in the design. The aim of this step is to achieve the desired throughput in the modified design [2]. This step is always followed by retiming via any commercial synthesis tool (see Fig. A.1), to translate the insertion of additional registers (via either re-pipelining or time interleaving) into an actual throughput improvement. This script is used in the flow associated with pipeline-level reconfiguration (shown in Fig. A.1), as well as the flow associated with thread-level reconfiguration (shown in Fig. A.2).

   **Inputs:**

   - *driven_wire_file:* The list of wires connected to outputs of all the flops in the design (generated by the *netlist2graph.py* script).
   - *driving_port_file:* The list of ports driving the wires in *driven_wire_file* (generated by the *netlist2graph.py* script).
   - *library:* Name of the library from which extra flip-flops have to be added.
   - *input_verilog_file:* Gate-level Verilog netlist (post-synthesis or Intellectual Property from third-party vendor).
   - *std_cell_technology:* The db format of the standard cell library (i.e., the timing library files that will be used by the synthesis and the place&route tool).
   - *no_of_pipeline:* Number of cascaded registers to be inserted from the input (i.e., additional pipestages).
   - *n:* Maximum number of threads in thread-level reconfiguration.
   - *comb_logic:* Flag determining whether the design to be re-pipelined is purely combinational. The flag is set to "true" if purely combinational, "false" if original input design contains sequential elements.

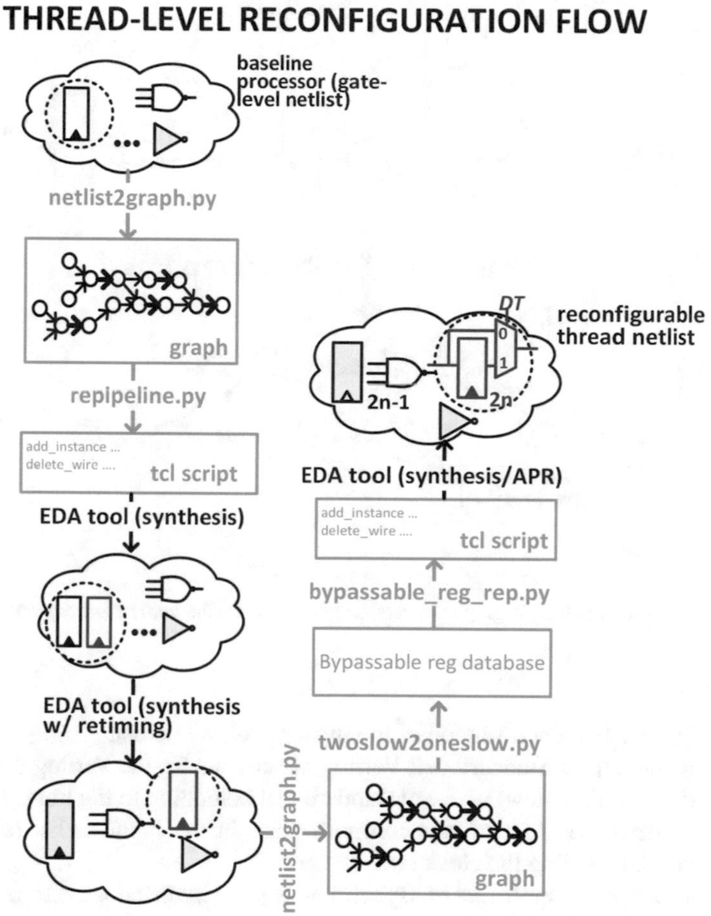

# THREAD-LEVEL RECONFIGURATION FLOW

**Fig. A.2** Reconfiguration flow detailing the usage of dedicated scripts for thread-level reconfiguration

**Outputs:**

- *dc_script:* Script to insert *"no_of_pipeline"* cascaded registers (i.e., pipe-stages) from the primary inputs, or replace each register by *"n"* cascaded registers for time interleaving.
- *output_verilog_file:* Verilog file for re-pipelined design.

**Run:** *python repipeline.py*

2. **netlist2graph.py:** This script converts the input structural gate-level Verilog netlist to the graph data structure in the form of associative array as shown in Fig. A.3. This script is used in the flow associated with pipeline-level (shown in Fig. A.1) and thread-level reconfiguration (shown in Fig. A.2).

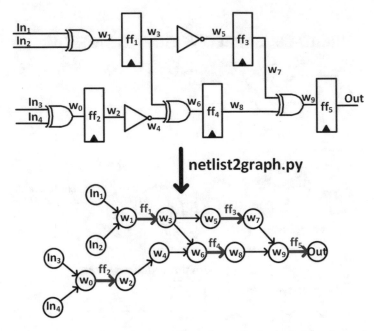

**Fig. A.3** Example of netlist-to-graph conversion performed by the *netlist2graph.py* script

**Inputs:**

- *ip_file:* Flattened version of the input gate-level netlist.
- *ip_tech_file:* Standard cell Verilog file containing the Verilog description (i.e., Verilog view) of every standard cell belonging to the library.
- *async_ports:* List of asynchronous ports in the input netlist (e.g., reset ports, including the clock itself).
- *async_ports_cell:* List of asynchronous ports in the adopted standard cells (e.g., set, reset, and the clock itself).
- *register_iden_pin:* Name of the clock port in flip-flop standard cells (e.g., CLK).

**Outputs:**

- *register_file:* File containing all the instance names of the flip-flops contained in the design.
- *ip_port_file:* File containing all the input ports of the input design (excluding *async_ports*).
- *op_port_file:* File containing all the output ports of the input design.
- *cell_ip_port_info_file:* File containing the input ports of each standard cells in the library.
- *cell_op_port_info_file:* File containing the list of output ports of all the standard cells in the adopted cell library. For example, a list item AND:Z refers to the output port Z of the AND cell, as extracted from the *ip_tech_file* file containing the Verilog view of all standard cells.

- *op_graph_file:* File containing the graph data structure in the following format:
  <wire_name1>, <wire_name2>=
  <port1>,<inst_name>,<mod_name>,<port_2>.
- *driven_wire_file:* List of all wires connected to the output of any flip-flop in the design.
- *driven_port_file:* List of ports driving the wires in *driven_wire_file*. This file contains the list of output ports and the flip-flops associated with it (e.g., FF4:Q refers to the output port Q of flip-flop #4).

**Run:** *python netlist2graph.py*

NOTE: By construction, it is necessary to restrict the types of flip-flops in the standard cell library (use *set_dont_use* feature in the synthesis tool while synthesizing the design) to basic D flip-flops with or without asynchronous and/or synchronous resets (i.e., exclude flip-flops with *enable* pins). Before running the *netlist2graph.py* script, it is necessary to ensure that the throughput requirement is met by the output design from the *repipeline.py* script. If the throughput requirement is not met, the number of inserted registers *no_of_pipeline* in *repipeline.py* needs to be increased.

3. **graph2skeletongraph.cpp:** This script converts the graph extracted from the re-pipelined netlist in Fig. A.1 to the corresponding skeleton graph containing the basic information about the flip-flops used in the design (see details in Sect. 3.4). This step is essential to reduce the size of the graph database to efficiently handle large designs. The combinational logic between any pair of launching and capturing flip-flops is reduced to a graph edge, as shown in Fig. A.4. This script is used in the flow associated with pipeline-level reconfiguration (shown in Fig. A.1).

**Inputs:**

- *ip_port_file:* File containing all the input ports in the input design (excluding *async_ports*).
- *op_port_file:* File containing all the output ports in the input design.
- *register_file:* File containing all the instance names of the flip-flops contained in the design.
- *nl_graph_file:* Output file containing the graph database created by the *netlist2graph.py* graph.

**Outputs:**

- *skl_graph_file:* File containing the output skeleton graph with the following format:
  <node$_1$ >, <node$_2$> = <edge_weight>, where
  <node$_i$> = input_wire: flip_flop_name: output_wire.
- *skl_graph_node_file:* File containing the list of the nodes of the skeleton graph.

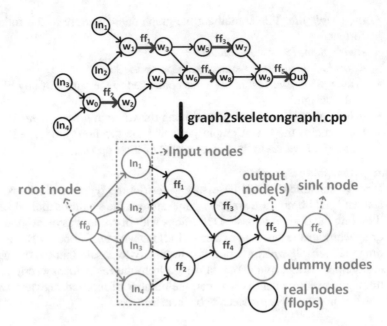

**Fig. A.4** Example of graph-to-skeleton graph conversion by the *graph2skeletongraph.cpp* script

- **dummy_nodes_file:** File containing the list of all the dummy nodes in the skeleton graph.

**Run:** *make graph2skeletongraph*

4. **weighskeletongraph.py:** This script fills the <edge_weight> field in the skeleton graph created after running *graph2skeletongraph.cpp*. The edge weight is defined and evaluated as the critical path delay between the launching and the capturing flip-flops represented by <node₁> and <node₂>, respectively. This file invokes the synthesis tool (Design Compiler in this specific script, easily changeable to any other tool), which populates the *edge_weight* field by using the file "*timing.tcl*" generated as an intermediate file in this step. See example in Fig. A.5. This script is used in the flow associated with pipeline-level reconfiguration (shown in Fig. A.1).

**Inputs:**

- **skeleton_graph_file:** File containing the skeleton graph with weights being initialized to "0," as created by *graph2skeletongraph.cpp*.
- **input_verilog_file:** Gate-level netlist of the original input design.
- **ip_port_file:** File containing all the input ports in the original input design.
- **op_port_file:** File containing all the output ports in the design.
- **register_file:** File containing all the instance names of the flip-flops contained in the design.

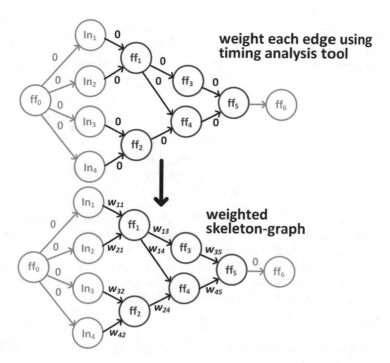

**Fig. A.5** Example of graph weighting by the *weighskeletongraph.py* script

**Outputs:**

- *skeleton_graph_file:* Graph associated with the input netlist updated with the weights.

**Run:** *python weighskeletongraph.py*

5. *identify_pipeline_linear.cpp:* This script performs register identification in linear pipelines [2], as summarized in Fig. A.6 (see details in Sect. 3.5). This script is used only when it is assured that the microarchitecture is purely linear (see Sect. 2.2) and only for pipeline-level reconfiguration (shown in Fig. A.1). In the opposite case of non-linear pipelines, the *identify_pipeline.cpp* script needs to be run.

**Inputs:**

- *skl_graph_file:* File containing the skeleton graph with the following format:
  $\langle node_1 \rangle$ , $\langle node_2 \rangle$ = $\langle edge\_weight \rangle$, where
  $\langle node_i \rangle$ = input_wire: flip_flop_name: output_wire.
- *skl_graph_node_file:* File containing the list of the skeleton graph nodes.
- *dummy_nodes_file:* File containing the list of dummy nodes in the skeleton graph.

**Fig. A.6** Example of register identification in linear pipelines performed by the *identify_pipeline_linear.cpp* script

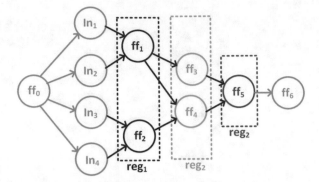

- *ip_port_file:* File containing all the input ports in the input design (or any other intermediate design, excepting the final one since it includes the extra *bypass* port to control the configuration).
- *op_port_file:* File containing all the output ports in the input design (or any other intermediate design, excepting the final one since it includes the extra *bypass* port to control the configuration).

**Outputs:**

- *registers_to_be_bypassed_file:* File containing the list of the registers to be bypassed.
- *register_level_file:* file containing the list of the registers and their corresponding pipeline number (which is unambiguously defined in linear pipelines, as discussed in Sect. 3.5).

**Run:** *make identifypipeline_linear*

6. *identify_pipeline.cpp:* This script performs register identification in non-linear pipelines, as summarized in Fig. A.1 (see details in Sect. 3.6). After identification, the script selects the most optimal set of registers to be bypassed, based on the algorithm described in Sect. 3.7. The script should be used only if the design contains loops and/or feedforward paths (see Sect. 2.2), as exemplified in Fig. A.7. This script is used in the flow associated with pipeline-level reconfiguration (shown in Fig. A.1).

**Inputs:**

- *skl_graph_file:* File containing the skeleton graph with the following format:
  <node$_1$ > , <node$_2$> = <edge_weight>, where
  <node$_i$> = input_wire: flip_flop_name: output_wire.
- *skl_graph_node_file:* File containing the list of the nodes of the graph, excluding the dummy nodes of the skeleton graph (see details in Sect. 3.4).
- *dummy_nodes_file:* File containing the list of all dummy nodes in the skeleton graph.

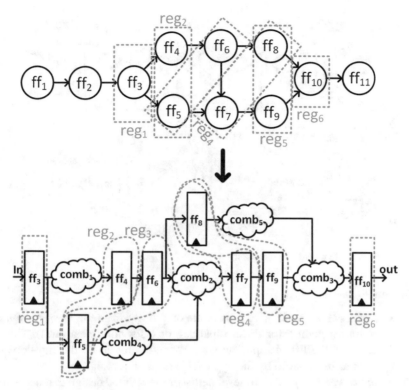

**Fig. A.7** Example of register identification in non-linear pipelines performed by the *identify_pipeline.cpp* script

**Outputs:**

- *set_of_cutset_file*: File containing all sets of cutsets present in the skeleton graph of the design. In short, this file gives all the sets of cutsets as described in Sect. 3.6.
- *loops_file*: File containing all the loops (i.e., the sequence of flip-flops present in any given loop) present in the design.
- *reg_to_be_bypassed_file:* File containing all the registers that have been identified to be the most optimal to be made bypassable.
- *reg_not_to_be_bypassed_file:* File containing the conventional registers remaining in the output design (i.e., not replaced by bypassable registers).

**Run:** *make identifypipeline*

7. *bypassable_reg_replacement.py:* This script replaces the registers to be made bypassable (as listed in *reg_to_be_bypassed_file*, see above) with the bypassable register version, as shown in Fig. A.8. This script is used in the flow associated with pipeline-level (shown in Fig. A.1) and thread-level reconfiguration (see Fig. A.2).

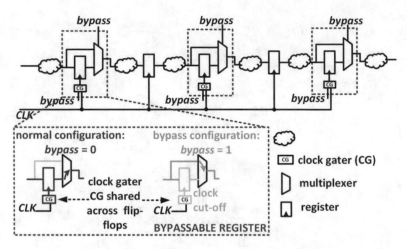

**Fig. A.8** Bypassable register replacement

**Inputs:**

- *replacement_node_file:* This is the output file from the *identify_pipeline* script (at point 5 for linear pipelines, or 6 for non-linear ones). This file contains the list of input/output ports and wires in all the flip-flops that need to be replaced by the bypassable register version.
- *input_verilog_file:* Gate-level netlist of the retimed design that needs to be reconfigured. This netlist is obtained after the execution of re-pipelining and retiming steps. The netlist will be later used for register identification.
- *std_cell_technology:* db (or lib) file used as link and target library in synthesis tool.
- *bypassale_flops_file:* Verilog gate-level netlist containing the basic flip-flop with a cascaded multiplexer to form the bypassable flip-flop configuration (which will replace the flip-flops to be made bypassable).
- *no_of_flops_per_gate:* Maximum number of flip-flops driven by each clock gater. This number should be set to a value above the expected number of bypassable flip-flops, so that only one clock gater is created and assigned to all flip-flops that are categorized as bypassable. The place&route tool will later optimize the number of flip-flops per clock gater, based on the clock load and wire load during clock tree synthesis (CTS) phase.

**Outputs:**

- *output_verilog_file:* Final output gate-level netlist of reconfigured design.

**Run:** *python bypassable_reg_replacement.py*

8. **twoslow2oneslow.py:** This script is used to identify the flip-flops that need to be replaced with bypassable registers in thread-level reconfigurable designs. This script is used in the flow associated with thread-level reconfiguration (shown in Fig. A.2).

**Input:**

- **graph_file:** File containing the graph data structure in Fig. A.2 in the following format:
  <wire_name1>,<wire_name2>=
  <port1>,<inst_name>,<mod_name>,<port_2>
  This graph is generated after passing the netlist through *netlist2graph.py* script as shown in Fig. A.2.
- **reg_file:** File containing all the flip-flop instances in the design.
- **ip_port_file:** File containing all the input ports in the design obtained after the re-pipelining (i.e., running *repipileing.py*) and the retiming step, as shown in Fig. A.2.

**Output:**

- **replacement_node_file:** File containing the registers that need to be replaced by bypassable registers.

**Run:** *python twoslow2oneslow.py*

# References

1. S. Jain, L. Lin, M. Alioto, Dynamically adaptable pipeline for energy-efficient microarchitectures under wide voltage scaling. IEEE J. Solid State Circuits 53(2), 632–641 (2018)
2. S. Jain, L. Lin, M. Alioto, Automated design of reconfigurable microarchitectures for accelerators under wide voltage scaling, in *IEEE Transactions on Very Large Scale Integration (VLSI) Systems* [in print]
3. S. Jain, L. Lin, M. Alioto, Drop-in energy-performance range extension in microcontrollers beyond VDD scaling, in *2019 IEEE Asian Solid-State Circuits Conference, Macau (China)*, (2019), pp. 125–128
4. S. Jain, M. Alioto, *RECMICRO: Design Framework and Scripts to Design Reconfigurable Microarchitectures* [Online], http://www.green-ic.org/recmicro
5. *NCSU Electronic Design Automation (EDA) Wiki* [Online], http://www.eda.ncsu.edu/wiki/NCSU_EDA_Wiki
6. N. Weaver, Y. Markovskiy, Y. Patel, J. Wawrzynek, Post-placement C-slow retiming for the Xilinx Virtex FPGA, in *FPGA*, (2003)

# Index

© Springer Nature Switzerland AG 2020
S. Jain et al., *Adaptive Digital Circuits for Power-Performance Range beyond
Wide Voltage Scaling*, https://doi.org/10.1007/978-3-030-38796-9